The Information Diet

Diet

A Case for Conscious Consumption

Clay A. Johnson

O'REILLY®

Beijing · Cambridge · Farnham · Köln · Sebastopol · Tokyo

The Information Diet
by Clay A. Johnson

Published by O'Reilly Media, Inc., 1005 Gravenstein Highway North, Sebastopol, CA 95472.

O'Reilly books may be purchased for educational, business, or sales promotional use. Online editions are also available for most titles (safari.oreilly.com). For more information, contact our corporate/institutional sales department: (800) 998-9938 or corporate@oreilly.com.

Editors: Julie Steele and Meghan Blanchette
Production Editor: Melanie Yarbrough
Proofreader: Melanie Yarbrough
Cover Designer: Mark Paglietti

Interior Designers: Ron Bilodeau and Edie Freedman
Cover Designer: Mark Paglietti
Illustrators: Robert Romano and Jessamyn Read

Printing History:

January 2012: First Edition.

Revision History for the First Edition:

2011-12-13: First Release

See *http://oreilly.com/catalog/errata.csp?isbn=9781449304683* for release details.

MAR 2012

ISBN: 978-1-449-30468-3

[CW]

To my dad, Ray Johnson. When he and my mom dropped me off for college, he told me that there were three lessons he'd learned from many decades of practicing psychiatry:

1. Don't jump in anybody else's drama
2. Always believe in yourself
3. Don't believe everything you think

Talk about a healthy information diet.

Contents

PART III: SOCIAL OBESITY 119

Preface

The things we know about food have a lot to teach us about how to have a healthy relationship with information. It turns out that foods that are bad for us have analogues in the world of information. In the world of agriculture, we now have factory farms churning out junk food; and in the world of media, we now have content farms churning out junk information. Consuming whole foods that come from the ground tends to be good for you, and consuming news from close to its source tends to inform you the most.

That's what this book is about. My hope is that by reading it, you will gain the knowledge and incentive to transform your relationship with information and have a healthier lifestyle as a result. You'll have more time to spend with your loved ones, be more effective at work, and be a more empowered citizen in your community.

For me, this book isn't just a book—it's a mission. Information overconsumption is a serious health problem for the American electorate, and we can see it from the halls of Congress to the tents of the Occupy Wall Street movement and the Tea Party. In any democratic nation with the freedom of speech, information can never be as strongly regulated by the public as our food, water, and air. Yet information is just as vital to our survival as the other three things we consume. That's why personal responsibility in an age of mostly free information is vital to individual and social health. If we want our communities and our democracies to thrive, we need a healthier information diet.

We'd Like to Hear from You

Please address comments and questions concerning this book to the publisher:

O'Reilly Media, Inc.
1005 Gravenstein Highway North
Sebastopol, CA 95472
(800) 998-9938 (in the United States or Canada)
(707) 829-0515 (international or local)
(707) 829-0104 (fax)

We have a web page for this book, where we list errata, examples, and any additional information. You can access this page at:

http://www.oreilly.com/catalog/9781449304683

To comment or ask technical questions about this book, send email to:

bookquestions@oreilly.com

For more information about our books, courses, conferences, and news, see our website at *http://www.oreilly.com.*

Find us on Facebook: *http://facebook.com/oreilly*

Follow us on Twitter: *http://twitter.com/oreillymedia*

Watch us on YouTube: *http://www.youtube.com/oreillymedia*

Safari® Books Online

Safari Books Online is an on-demand digital library that lets you easily search over 7,500 technology and creative reference books and videos to find the answers you need quickly.

With a subscription, you can read any page and watch any video from our library online. Read books on your cell phone and mobile devices. Access new titles before they are available for print, and get exclusive access to manuscripts in development and post feedback for the authors. Copy and paste code samples, organize your favorites, download chapters, bookmark key sections, create notes, print out pages, and benefit from tons of other time-saving features.

O'Reilly Media has uploaded this book to the Safari Books Online service. To have full digital access to this book and others on similar topics from O'Reilly and other publishers, sign up for free at *http://my.safaribooksonline.com.*

Acknowledgments

I'd first like to thank my incredible wife, Rosalyn Lemieux. She's been a valuable sounding board for this book, helped me clarify some of my own ideas, and probably read more drafts than anybody. And she was a good sport in allowing me to expose Zombie Roz and Email Roz.

Both my parents, Joy and Ray Johnson, are remarkable, and their input into this book, over more than three decades, should not go unnoticed either.

My editors, Julie Steele and Meghan Blanchette, are also magnificent. They've made this book not a weird rambling of strange ideas, but a cogent story. When I first started the journey of writing this book, I thought about self-publishing, but the value of strong, smart editors still justifies the existence of publishers.

Rebecca Bell was also instrumental in the writing of this book by allowing me to use her wonderful home off the coast of Georgia to escape the distractions of high bandwidth and focus entirely on writing. I could not have done it without that amazing gift.

My network of colleagues, friends, and family has been indispensable throughout the creation of this book: Jen Pahlka, Carl Malamud, Howard Rheingold, Anil Dash, Andy Baio, Noreen Nielsen, Karl Frisch, Eric Burns, Jake Brewer, Mary Katharine Ham, Michael Bassik, Tom Hughes-Croucher, Pete Skomoroch, Jane McGonigal, Jim Gilliam, Josh Hendler, Cammie Croft, Steve Geer, Tom Steinberg, Mario Flores, Cindy Mottershead, Maggie McEnerny, Todd Kamin, and Cheryl Contee—thank you so much for sitting and listening to me describe my book and pushing me when I needed it. Without you all, this book would not have happened.

The countless people I interviewed on and off the record—thank you. You know who you are. Linda Stone, you are a national treasure. Anybody who has read anything in this book and gotten anything out of it ought to listen to what she has to say.

The people who helped me edit this book were saviors. Eric Newton took the time to send me amazing feedback and provide me more historical context than I could ever ask for. Quinn Norton's brutal honesty helped sharpen my focus and my argument, and without Clay Shirky and Gina Trapani's encouragement, this book probably would have never seen the light of day.

Finally, a tip of the hat ought to go to Tim O'Reilly for giving me a platform to share this with you. He's a mentor and a friend who does not get enough credit for injecting his community with the right kinds of values. Thank you, Tim.

Introduction

"When you're young, you look at television and think, There's a conspiracy. The networks have conspired to dumb us down. But when you get a little older, you realize that's not true. The networks are in business to give people exactly what they want. That's a far more depressing thought. Conspiracy is optimistic! You can shoot the bastards! We can have a revolution! But the networks are really in business to give people what they want. It's the truth."

—Steve Jobs[1]

When I saw the cardboard sign—which displayed what had to be the craziest seven words I'd seen in a long time—I knew I had to quit my job.

I was working for the Sunlight Foundation, a government transparency operation in Washington, D.C. The premise of the organization was simple: if we give people access to government data, they will demand better government, they will vote differently, and the quality of politicians getting elected will improve. But these seven words, held above the head of what looked to be a 40-something male in front of the White House, broke my heart and made me realize how futile that mission was by itself.

The sign said: "Keep your government hands off my Medicare."

But I'm getting ahead of myself.

I'd spent the past 10 years in Washington, D.C., trying to make a difference. Lots of folks said that my call to politics was like a call to the priesthood: that I was meant for it. It started in 2000, when I did get a call, but it wasn't from God or even from Washington, D.C.—it was from my mother.

---

1   *http://www.wired.com/wired/archive/4.02/jobs.html*

She told me the doctors had found a lump in her breast, and that the diagnosis was breast cancer. Because my father had retired, my mother was ineligible for Medicare and was independently insured; her monthly insurance premiums were going to go from $300 a month to $3,000 a month. My father would have to come out of retirement to work as a psychiatrist for the state of Georgia so that my mother could undergo therapy and have affordable access to healthcare.

I, the bright-eyed twenty-something, thought I'd do something about it, and a couple of years and a war in Iraq later, I found myself driving up to Burlington, Vermont, to work for Governor Howard Dean. I'd never voted before, but Dean was a medical doctor, and had made reasonable moves on healthcare in his home state. He was the only person running for president who seemed like he could get my mother's health insurance premiums down, and make it so my dad could retire again.

That started my career in politics. After the Dean Campaign ended, I was still convinced that electing Democrats would help get my mom's health insurance premium down, so I went on to cofound a company called Blue State Digital with three of my friends from the Dean Campaign. Our plan was to take the lessons and technology we'd learned and turn them into a business that would help the Democrats raise money and win votes over the Internet. My naive belief at the time was that if we could simply elect more Democrats to Congress and to the White House, then my mom's health insurance problem would get fixed.

The story is almost a cliché from there. The company was very successful. I was making far more money than I'd ever made before. But it became obvious (about four years later) that I wasn't solving the problem that I'd set out to solve. After electing a majority of Democrats to the House in 2006 and still seeing no movement on healthcare, I decided that electing Democrats to fix the problem wasn't doing a whole lot of good. There must have been some other impediment I needed to address.

Lobbyists! Of course, it was the lobbyists—those dark evil characters in the backs of high-end, smoke-filled cigar bars in Washington, bribing our members of Congress to vote against the will of the American people. Surely it was them.

I left the company after watching Barack Obama, soon to become the nation's best-known client of Blue State Digital, win the Iowa Caucuses in the winter of 2008. My new job at the Sunlight Foundation was directing a squad of technologists. Our mission was to liberate and analyze government data, and to make it easier for people to make more informed decisions about elections. If we could show America with hard facts that their Congress was being bought off, surely that would spur them to action.

After two years on the job at Sunlight—a full eight years since my mom was diagnosed and two radical mastectomies later—I watched the newly-elected President Barack Obama bring up healthcare. It should have been a great moment, the realization of my hopes for nearly a decade. Instead, I watched the nation go into a bitter and angry debate about the role of our government. Ironically, this was about the same time that my mom became eligible for Medicare.

The news media was saturated with every kind of graph and chart about our healthcare costs, wait times, the efficiency of government, how Canada does it, how old people handle healthcare, and what kinds of medicines would and would not be available to Americans should we pass some form of healthcare overhaul. At Sunlight, we did our best to stick to the facts. We built "Sunlight Live," which allowed people to watch the healthcare debates online; next to each member of Congress when they spoke appeared the amount of money they had received from the healthcare industry.

During that long, bitter, and angry debate, I took a stroll down to the White House. And that's when I saw that sign, those jarring seven words, held high:

"Keep your government hands off my Medicare."

It's amazing how the little things can give you perspective. But then I spoke to this protestor about his sign. He seemed rather well educated—sure, he was angry, but he was not dumb, just concerned about the amount of money being spent by the current administration. He talked to me about topics that I, as a professional in Washington for 10 years, hadn't really thought about since my political science classes in college. This man did not suffer from a lack of information. Yet he had failed to consider the irony of holding a sign above his head asking government to keep its hands off a government-run program. To him, it made perfect sense.

Then something else happened. I live near Walter Reed Hospital, a hospital that treats injured veterans. On my jogs around my block, Marines—likely injured from our military operations—often breeze by me with one leg and one prosthetic. Although it certainly wounds my self-esteem as a runner, it's a miracle. Yet at the front of the gates to the hospital on Georgia Avenue one evening, I spotted another sign:

"Enlist Here To Die for Halliburton."

I imagined that sign-maker, Halliburton Woman, to be the polar opposite of Medicare Man. But equally wrong. I mean, even on the surface: nobody enlists for the United States Army at the Army *hospital*.

How could it be possible that educated, intelligent people have somehow become capable of believing in a distorted reality? At that moment, an idea popped into my head. What if a person's native or learned abilities to process information sensibly could be warped by feeding junk into the mental machine? As we say in technology: *garbage in, garbage out.*

We know we're products of the food we eat. Why wouldn't we also be products of the information we consume?

In the case of the activists—one against Obama's healthcare plan, the other against our wars in Afghanistan and Iraq—these people certainly seemed *informed*. But it's as though they had caught some kind of disease that made it so they couldn't think clearly.

If unhealthy information consumption creates bad information habits the way unhealthy eating creates food addictions, then what good is transparency? I left the Sunlight Foundation. Transparency wasn't the universal answer I was looking for. You cannot simply flood the market with broccoli and hope that people stop eating french fries. If large numbers of people only seek out information that confirms their beliefs, then flooding the market with data from and about the government will really not work as well as the theorists predict; the data ends up being twisted by the left- and right-wing noise machines, and turned into more fodder to keep America spinning.

Today, you're likely to spend upwards of 11 hours per day consuming information—reading books like this, checking out your friends' Facebook pages, reading the newspaper, watching television, listening to the radio or your portable music player. For many of us who work in front of a computer all day, it's even more: we spend all day reading and writing in front of a screen.

The sheer amount of information available to us is mind-boggling. According to storage company EMC, there are presently 800,000 petabytes (each petabyte representing one million gigabytes) in the storage universe. And according to the University of California in San Diego, American homes consume nearly 3.6 zettabytes (one million petabytes are in one zettabyte) of information per day. It's expected to grow, too: EMC expects a 44-fold increase in data storage by 2020.

So we've come up with this term to deal with it: *information overload*. (If you search for "information overload" on Amazon.com, you'll get 9,093 results—roughly eight and a half times more than the number of results that a search for "irony" returns.)

For a new professional class, achieving "inbox zero" (dealing with every email inside of your inbox) is akin to running a 10k or getting a promotion at work. We've also developed a near bogey man type of mythology around our information abundance. In 2011, Nicholas Carr was nominated for a

Pulitzer Prize for his book *The Shallows: How the Internet Is Rewiring Our Brains* (W.W. Norton).

Using Google's n-gram viewer—a service Google provides that allows you to count how many times a phrase appears in its giant corpus of books over 150 years—you can see that the term information overload became popular just after 1960, and surged 50% by 1980 and and again by 2000.

The concept of information overload doesn't work, however, because as much as we'd like to equate our brains with iPods or hard drives, human beings are biological creatures, not mechanical ones. Our brains are as finite in capacity as our waistlines. While people may eat themselves into a heart attack, they don't actually die of overconsumption: we don't see many people taking their last bite at a fried chicken restaurant, overstepping their maximum capacity, and exploding. Nobody has a maximum amount of storage for fat, and it's unlikely that we have a maximum capacity for knowledge.

Yet we seem to want to solve the problem mechanically. Turn it the other way around and you see how absurd it is. Trying to deal with our relationship with information as though we are somehow digital machines is like trying to upgrade our computers by sitting them in fertilizer. We're looking at the problem through the wrong lens.

Instead of the lens of efficiency and productivity, maybe we should start looking at this through the lens we use to view everything else we biologically consume: health.

What if we started managing our information consumption like we managed our food consumption? The world of food consumption and the world of information consumption aren't that far apart: both the fields of cognitive psychology and neuroscience show us that information can have physiological effects on our bodies, as well as fairly severe and uncontrollable consequences on our decision-making capability.

When viewed through this lens, the information abundance problem appears more dire. Coping with the problem isn't a matter of getting things done anymore; it's a matter of health and survival. Information and power are inherently related. Our ability to process and communicate information is as much an evolutionary advantage as our opposable thumbs.

There are kinds of food we're hard wired to love. Salt, sugars, and fats. Food that, over the course of the history of our species, has helped us get through some long winters, and plow through some extreme migrations. There are also certain kinds of information we're hard wired to love: affirmation is something we all enjoy receiving, and the confirmation of our beliefs helps us form stronger communities. The spread of fear and its companion, hate, are clearly survival instincts, but more benign acts like gossip also help us spread the word about things that could be a danger to us.

In the world of food, we've seen massive efficiencies leveraged by massive corporations that have driven the cost of a calorie down so low that now obesity is more of a threat than famine. Those same kinds of efficiencies are now transforming our information supply: we've learned how to produce and distribute information in a nearly free manner.

The parallels between what's happened to our food and what's happened to our information are striking. Driven by a desire for more profits, and a desire to feed more people, manufacturers figured out how to make food really cheap; and the stuff that's the worst for us tends to be the cheapest to make. As a result, a healthy diet—knowing what to consume and what to avoid—has gone from being a luxury to mandatory for our longevity.

Just as food companies learned that if they want to sell a lot of cheap calories, they should pack them with salt, fat, and sugar—the stuff that people crave—media companies learned that affirmation sells a lot better than information. Who wants to hear the truth when they can hear that they're right?

Because of the inherent social nature of information, the consequences of these new efficiencies are far more dramatic than even the consequence of physical obesity. Our information habits go beyond affecting the individual. They have serious social consequences.

Much as a poor diet gives us a variety of diseases, poor information diets give us new forms of ignorance—ignorance that comes not from a lack of information, but from overconsumption of it, and sicknesses and delusions that don't affect the underinformed but the hyperinformed and the well educated.

Driven by a desire for more profits, and for wider audiences, our media companies look to produce information as cheaply as possible. As a result, they provide affirmation and sensationalism over balanced information. And in return, we need to start formulating an information diet—what to consume and what to avoid—in this new world of information abundance.

The first step is realizing that there is a choice involved. As much as our televisions, radios, and movie theaters would have us believe otherwise, information consumption is as active an experience as eating, and in order for us to live healthy lives, we must move our information consumption habits from the passive background of channel surfing into the foreground of conscious selection.

The first part of this book is intended to give you a good idea of how we got to where we are—to explore the economics of information, and the biological consequences of our information consumption.

The second part of the book is an attempt to design an information diet—describing the healthy habits of a good information consumer, and providing pointers on how to consume that information.

The third part of this book is a call to action: if our information consumption has a social consequence, then it's not only about ourselves, but also about ethics. Just as the food we eat has an ethical consequence, so do the choices we make around information. In order to create better access to information, better quality sources, and healthier lifestyles, suppliers must change. And suppliers will only change with proven demand. If things are to truly change, then we've got to break the insidious cycle that we ordinary people create with our demand, and media companies create with more supply.

This book is also the outcome of my experience as a political operative and transparency advocate in Washington, D.C. It discusses politics, and while I do draw from my experiences as a consultant to Democratic campaigns and causes, I attempt to be even-handed with my discussion. My goal is not to convince you to be a liberal, or a conservative, but rather to close the gap between you and your government by giving you some insight about what is really going on in Washington.

Like any good diet, the information diet works best if you think about it not as denying yourself information, but as consuming more of the *right stuff and developing healthy habits*. The result I wish for you is just what you'd expect from any kind of balanced diet: a healthier and happier lifestyle.

The more I researched the parallels between our food consumption and our information consumption, the more strongly I came to believe that this isn't just a fancy metaphor. It's real. Conscious consumption of information is possible. We can (and some already do) pay as much attention to the information we put into our heads as we do to the food we put into our bodies. Much like a healthy food diet, a healthy information diet has consequences that not only can reduce stress but also may help us live longer, happier lives.

CHAPTER 1

Lessons from Obesity

"What we know about diets hasn't changed. It still makes sense to eat lots of fruits and vegetables, balance calories from other foods, and keep calories under control. That, however, does not make front-page news."

—Marion Nestle, Food Activist[2]

William Banting learned the hard way that you are what you eat, and as a result, he invented what we know today as the modern diet.

An undertaker from Great Britain, Banting found himself suffering from "failing sight and hearing, an umbilical rupture requiring a truss, and bandages for weak knees and ankles." He reported not being able to walk down stairs without help, or to touch his toes. He went to see many doctors for his various conditions but claimed that, "not one of them pointed out the real cause of my sufferings, nor proposed any effectual remedy." The real cause of Banting's suffering wasn't that he couldn't walk down stairs, it was that he was obese.

After he started losing his hearing, he finally sought specialized medical attention and found himself in the care of "the celebrated aurist" Dr. William Harvey. The physician put him on a diet inspired by a lecture he'd heard about treating diabetes: five to six ounces of meat or fish three times a day, accompanied by stale toast with cooked fruit. Beer, potatoes, milk, and sweets were not allowed. Alcohol was, though: four to five glasses of wine a day, a glass of brandy in the evening, and sometimes even a wake-up cocktail in the morning were called for.

Banting reported losing 13 inches off his waist and 50 pounds of weight over the course of a couple of years. It was only then that Banting realized

---

2 http://www.althealth.co.uk/news/latest-news/diet-study-confusion-will-not-change-habits-analysts/

that he had been treating symptoms, not the root cause. Once he fixed his diet, his other problems went away. He could walk down the stairs again.

We've known that obesity is bad for a very long time. In the fourth century BCE, Hippocrates, called the father of medicine by Western scholars, wrote, "Corpulence is not only a disease itself, but the harbinger of others." And the Bible is filled with warnings about overconsumption. Proverbs 23:20–21 says, "Be not among winebibbers; among riotous eaters of flesh: For the drunkard and the glutton shall come to poverty: and drowsiness shall clothe a man with rags."

However, for thousands of years, obesity was usually a disease affecting only the most affluent. Food—especially the delicious, calorie-dense stuff—was simply too expensive for the average person to obtain. Few could afford to be fat, and thus being so was often considered a way to display one's prosperity.

Then a great technological shift happened, much like the one that we faced in the second half of the twentieth century. New technology and new techniques increased our food supply. The steam engine, crop rotation, and the iron plow revolutionized agriculture in Europe between the 17th and 19th centuries, alongside a variety of sociopolitical changes, including the rise of the merchant class. The food supply became more abundant, and access to it improved. Obesity was no longer just for a fortunate few.

It was in this context that Banting decided to share his results with the world. In 1863, he published Europe's first modern diet book, *Letter on Corpulence*, and sold an astounding 63,000 copies for a shilling each. It was the first diet craze of the West (called, appropriately, *banting*), and thousands were inspired to lose weight with his diet. The book also had global reach. It was translated into multiple languages and according to Banting, achieved good sales in France, Germany, and the United States.

The medical community treated it as old news. Their critique wasn't an assault on the idea, but they questioned why Banting's letter was so popular in the first place. Similar works had been published prior to his, but they were written by physicians, for physicians. *Letter on Corpulence* was written by a suffering person, for suffering people. His message resonated. People were ready to hear it. And Banting provided it in a form they could understand.

In the fourth edition of his letter, Banting spends upwards of seven pages defending himself against a medical fraternity that disputed his story, claiming that he must not have sought the attention of particularly good doctors if it took him that long to get well, or worse, that Banting's recommendation of four meals a day would cause more corpulence. His response:

> "My unpretending letter on Corpulence has at least brought all these facts to the surface for public examination, and they have thereby had already a great share of attention, and will doubtless receive much more until the system is thoroughly understood and properly appreciated by every thinking man and woman in the civilized world."

A Modern Epidemic

Banting was right about all the public attention—the commercial success of his pamphlet helped create an industry of diet books, coaches, and consultants. His documents are preserved online by the Atkins Foundation, the organization dedicated to Robert Atkins, who would come along more than a century later and encourage people to go on a very similar low-carbohydrate program.

But neither Banting nor Atkins, nor any of the thousands of others, solved the problem of obesity. In recent years, it has run rampant through America.

The Centers for Disease Control in Atlanta provides annual data on a state-by-state basis regarding our obesity epidemic. Figure 1-1 shows what obesity looked like in 1990.

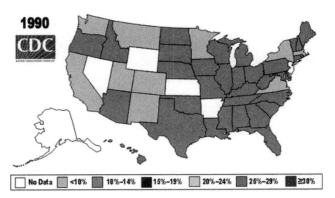

Figure 1-1. A map showing annual obesity rates for each state in 1990.

While the CDC does not have data for five states in 1990, none of the states for which data was collected had obesity percentages higher than 14%. Figure 1-2 shows what that same map looks like with data from 2010.

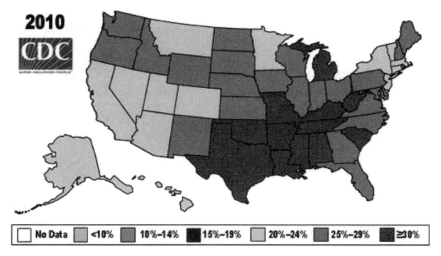

Figure 1-2. A map showing annual obesity rates for each state in 2010.

In 20 years, we went from an obesity rate no higher than 14% in any state to an obesity rate no lower than 20% in any state, and an obesity rate higher than 25% in most states. Twelve states—Alabama, Arkansas, Kentucky, Louisiana, Michigan, Mississippi, Missouri, Oklahoma, South Carolina, Tennessee, Texas, and West Virginia—now have obesity rates greater than 30%. Moreover, our obesity rate is accelerating; we're getting fatter faster than we were 20 years ago.

What has happened?

The same things that have always happened. Food is cheaper. We can afford more of it. And we increased the number of steps between our food's source and our bellies so much so that our food doesn't even look like food anymore.

To start with, calorie-dense foods are now less expensive and more readily available than ever. According to the USDA, we're now producing 3,800 calories per person per day. That number is an increase of several hundred calories since 1970. And accordingly, 62% of adult Americans are now overweight, according to the National Center for Health statistics—in 1980, that number was 46%.[3]

---

3   *http://www.usda.gov/factbook/chapter2.pdf*

The Birth of Industrial Agriculture

In the twentieth century, agriculture went through profound changes, both in the United States and globally. At the beginning of the twentieth century, the rural farmer was the largest demographic in the United States. Nearly a century ago, more than 50% of the United States population lived in rural areas, and farming represented 41% of the American workforce.[4]

Then, industrialization happened. The development of the Ford Model-T, the tractor, pesticides, and other agricultural technologies brought a new drive for efficiency into America's heartland.

It took a century to double food production from 1820 levels to those in 1920. It took just 30 years to double it again, between 1920 and 1950. It took 15 years from 1950 to 1965, and 10 between 1965 and 1975. Food production has continued to grow exponentially as science and the demand for food has caused our agricultural industry to industrialize.[5]

In the words of food activist Michael Pollan:

"In the past century American farmers were given the assignment to produce lots of calories cheaply, and they did. They became the most productive humans on earth. A single farmer in Iowa could feed 150 of his neighbors. That is a true modern miracle."[6]

This drive and industrialization is necessary, actually. By 2050, the UN estimates, we'll need to double our food production again to maintain projected population growth.[7]

The miracle of abundance comes with a remarkable set of consequences. Today, America's heartland is empty; only 17% of Americans live in rural communities. Efficiency also means fewer jobs: if a single farmer can feed 150 neighbors, it means you need fewer farmers. Today, less than 2% of the United States population is directly employed in agriculture.

Another significant consequence of industrialization is a rise in occupational health hazards. Agriculture is now one of the most dangerous professions in America. According to the Centers for Disease Control,[8] agriculture is dominated by giant factory farms with livestock packed at huge scale. It's what allows just four companies to produce 81% of the cows, 73% of the

---

4   *http://www.ers.usda.gov/publications/eib3/eib3.htm*

5   Scully, Matthew. *Dominion* (p. 29). St. Martin's Griffin: 2003.

6   *http://longnow.org/seminars/02009/may/05/deep-agriculture/*

7   *http://www.un.org/News/Press/docs/2009/gaef3242.doc.htm*

8   *http://www.cdc.gov/niosh/topics/agriculture/*

sheep, 57% of the pigs, and 50% of the chickens in America.[9] But with a relatively homogenized and industrialized production system, toxins travel faster in our food supply and affect more people. Packing animals in makes them more susceptible to disease, which means that more antibiotics and other drugs go into our livestock, decreasing their effectiveness for treating human diseases. And locating these farms near places where produce is grown means an increased risk of food contamination.

While calories have become more affordable, the nutrients in our food have slowly disappeared before our eyes, only to be replaced with corn-based sugar, soy-based fat and protein, and a whole lot of salt. Sugar, corn, and soy are the three crops of America, and they are the crops that make it into our animals and onto our dinner tables. Our grocery stores are full of manufactured foods, made mostly from corn and soy, that aren't particularly good for us and as Michael Pollan says, not something that your grandmother would recognize as food.

We don't grow our food anymore; we manufacture it. And in the 1970s, we eliminated requirements for calling artificial food "artificial." So right on the front of the packages, in really big emotion-grabbing type, companies could create exciting new forms of false food without ever saying so. Add to that the fact that this cheaper, more highly processed food alters our taste sensation so that the organic stuff seems bland, and it's easy to see why we've gotten obese. Figures 2-1 and 2-2 speak for themselves.

A New Set of Consequences

As probability would have it, I am one of the 62% of Americans considered overweight. In fact, according to my Body Mass Index (BMI), I'm one of the 27% of us considered obese. It's a terrible feeling, compounded by my constant attempts at exercise. I've run a marathon and several half marathons; gone through the constant physical, mental, and intellectual abuse of Tony Horton's P90x; and endured the torture of vinyasa yoga. My house is littered with high-tech gadgetry—from the Wii Fit to the Withings Internet-powered scale (it posts your weight on Twitter) to the FitBit (a little gadget you wear on your belt to tell you how many calories you've burned by walking around). I do all of this for two reasons: so that I don't need to grease the doorframes to get out of the house, and so that I can eat.

If you want to know why Americans are getting fat, ask a fat person. It's because for most of us, food—especially food that's bad for us—is delicious.

---

9   Testimony by Leland Swenson, president of the U.S. National Farmers' Union, before the House Judiciary Committee, September 12, 2000.

Let's be clear: a chocolate-chip cookie, for most people, is superior in every way to a head of lettuce, and if one could make a chocolate-chip cookie as good for you as a pile of spinach, the salad industry would be in danger of being obliterated by Mrs. Fields. It's a cruel joke that the stuff that tastes the best is often the stuff that's worst for us. But the reason behind it makes a lot of sense.

From way back during mankind's foraging days until just a few centuries ago (less than a blink through the eyes of human history), we didn't have a lot of food to go around. We never knew where our next meal would come from, and thus our bodies became wired for scarcity. Over the millennia, we evolved into energy conservation machines. It's why we crave that salt-fat-sugar combination—and why, as far as I'm personally concerned, the most dangerous place in America is between me and a chicken wing.

Our bodies are programmed to acquire as many resources as possible, and to take what we don't need and store it as fat—fat that will keep us warm and supply us with energy during the harsh winter when there's a lot less out there to eat.

For modern society, neither winter nor famine is the same threat they used to be. Not only can you get "fresh" tomatoes at your grocery store in the middle of winter in the coldest regions of America, the season has all but stopped killing us. In 2010, according to the U.S. Natural Hazard Statistics report, there were only 42 deaths from "winter" in the United States and another 34 from cold. You have about as much chance of dying from the cold as you do from lightning.[10]

The fresh, warm Krispy Kreme donut on a cool fall morning does more than treat (and trick) the tastebuds. That sort of food production allows the planet to sustain ever-larger populations—we've now surpassed seven billion human beings—with cheap calories.

Once the biggest threats America faced were the Famed Four Horsemen of the Apocalypse—war, famine, pestilence, and disease. But we've traded them in for a new killer. Today, 13.5 million[11] people die each year of heart disease and stroke, and 4 million from diabetes-related complications[12]— far more than die in automobile accidents. Heart disease is now our number one killer, and it takes more people to the grave in the United States in five years than all our war-related deaths combined. Instead of dying from the cold of winter, we find death in cholesterol.

---

10 http://www.weather.gov/om/hazstats.shtml

11 http://www.who.int/mediacentre/factsheets/fs310/en/index2.html

12 http://www.worlddiabetesfoundation.org/composite-976.htm

The Modern Diet

Today, nearly a century and a half after Banting's diet recommendations, Amazon.com lists nearly 20,000 books available for purchase on diets.

As of this writing, 681 books have been released in the past 30 days, with another 112 coming soon. In the first six months of 2011, more than 2,000 books on weight-loss were released.

We know beyond any scientific doubt that being fat kills. For more than a century and a half, the medical community has known that sugars and carbohydrates make us fat. But we keep eating, despite all of the information and transparency—from this month's 681 diet books, to the nutritional labels on every item in the grocery store, to chains of Weight Watchers and Jenny Craigs across the country.

Food's scrumptious properties aside, perhaps the diet books have something to do with our obesity problem. It certainly seems as though the number of diet books available to the public correlates to obesity rates. While any economist would predict that the number of books to help people be less fat would grow with the number of fat people in the market, at what point does causation flow the other way around? Maybe all these diet books are making us fat by making it harder to figure out what a healthy diet is. At the very least, the modern obsession with weight over diet brings us a significant health issue. Imagery of being perfectly shaped and skinny trumps being healthy and happy, and as a result, scores of people suffer— and die—from eating disorders.

No matter which way you turn, abundant information makes it easy to distort our relationship with food into something unhealthy. If you're looking to surf through a land of false promises, spend a few minutes in the diet aisle of your local bookstore. You can lose weight by thinking like either a caveman or a French woman, or by eating only food that's cooked slowly. You can lose it, says the updated 2012 edition of *Eat This Not That!* (Rodale Books), by simply swapping in a Big Mac® for a Whopper-with-cheese®.

In the diet aisle, our relationship to food can take on social, political, and environmental significance. A healthy diet mustn't just include the right number of calories and the right interaction of nutritional elements. It must also produce the least amount of carbon, and be as natural as possible. It's no longer good enough to eat reasonable portions of lean meat; the meat must come from a cow that could roam free and eat grass.

If time is of the essence, you can get top-selling weight loss books promising change based on your lifestyle: just spend 8 minutes (*Eight Minutes in the*

Morning, Harper), 4 hours (*The Four Hour Body*, Crown Archetype), or 17 days (*The Seventeen Day Diet*, Free Press). If religion or the supernatural is your thing, just look to the 159 diet books available containing the word "miracle."

Most of what these books cover and the pseudoscience behind them appeals to the same emotional impulses as do the people peddling calories in the first place. Some of this is unavoidable in a free society: the right answers—healthy information—compete side-by-side with the answers we may want to hear but which may not be true. Only the highly nutritionally literate can easily tell the difference.

The best food journalists distill this complex world of choices into healthy ones. Michael Pollan, Knight Professor at the University of California at Berkeley, is a leading example. The beginning of his *In Defense of Food* (Penguin) is a seven-word diet guide: "Eat food. Not too much. Mostly plants." And there it is, right up front. We can take those three simple rules—those seven words—into the grocery store, and win.

While our collective sweet tooth used to serve us well, in the land of abundance it's killing us. As it turns out, the same thing has happened with information. The economics of news have changed and shifted, and we've moved from a land of scarcity into a land of abundance. And though we are wired to consume—it's been a key to our survival—our sweet tooth for information is no longer serving us well. Surprisingly, it too is killing us.

CHAPTER 2

Information, Power, and Survival

Most of us don't remember the moment we began to discover language, to understand words, and to speak—the closest we get to it is watching our children discover it for themselves.

Helen Keller was an exception. The renowned deaf-blind activist didn't learn about language until she was seven years old. Of the discovery, she wrote:

> "We walked down the path to the well-house, attracted by the
> fragrance of the honeysuckle with which it was covered. Some one
> was drawing water and my teacher placed my hand under the spout.
> As the cool stream gushed over one hand she spelled into the other
> the word water, first slowly, then rapidly. I stood still, my whole
> attention fixed upon the motions of her fingers. Suddenly I felt a
> misty consciousness as of something forgotten—a thrill of returning
> thought; and somehow the mystery of language was revealed to me.
> I knew then that "w-a-t-e-r" meant the wonderful cool something
> that was flowing over my hand. That living word awakened my
> soul, gave it light, hope, joy, set it free! There were barriers still,
> it is true, but barriers that could in time be swept away. I left the
> well-house eager to learn. Everything had a name, and each name
> gave birth to a new thought. As we returned to the house every
> object which I touched seemed to quiver with life. That was because
> I saw everything with the strange, new sight that had come to me."[13]

Keller discovered not just language, but a passion to know, and to tell—a sudden connection to the entire world. She describes the moment as gaining consciousness itself. The moment she began to understand that things had names, she said, was the moment she really began to *think*.

The origins of language aren't yet definitive. But many scientists agree that sometime between 50,000 and 200,000 years ago, our brains began to

---

13 http://www.afb.org/mylife/book.asp?ch=P1Ch4

change. We don't know whether this physiological change was viral (co-evolving with language, spreading from homo sapiens to homo sapiens at a rapid rate), or evolutionary (developing over thousands of years), or a combination of slow evolution turning suddenly into algorithmic adoption. Yet one thing is certain: once we had the capacity to communicate complex forms of information, we had a huge advantage over other species.

The fossil record tells us that when our species discovered language—and thus discovered complex forms of information—human society transformed. Around the same time language was developed, humans began leaving the continent of Africa in droves. This development of language, some scientists suggest,[14] is what enabled humans to organize, move together, and explore beyond the African continent.

Most importantly, at this moment, our species began to become more aware of ourselves. While it's evident that the mind co-evolved with language, it was at this moment that we were not only able to communicate increasingly complex concepts to one another, but also to store it in our brains effectively. Our cognition advanced.

Knowledge Is Power

With language came the ability to coordinate with each other more effectively. Nomadic tribes began to develop symbols to keep themselves better organized. Calendars appear to have been developed about 10,000 years ago, improving our ability to plant seasonal crops. Armed with the seeds of agriculture, we didn't need to be as transient anymore; gradually, supported by the surpluses of expanding agriculture, nomadic tribes turned into civilizations, and more sophisticated governments emerged.

The Sumerians and then the Egyptians started using glyphs to express the value of currency around 6,000 years ago. Once the Egyptians settled on a standard alphabet years later, they reaped another information technology boom and reached heights no other civilization previously known to man ever had, taking on massive engineering tasks, building new modes of transportation, and acquiring vast power across an empire.

Yet carving symbols into stone tablets was painstaking work, and errors were costly. Stone tablets didn't travel that well either, and if they broke, weeks, months, or even years of work could be lost in an instant. As such, production of this kind of information was largely relegated to a special class: scribes.

---

14 *http://www.accessexcellence.org/BF/bf02/klein/bf02e3.php*

Being a good scribe meant holding significant power in Egypt.[15] Not only were scribes exempt from the manual labor of the lower classes in Egypt, but many also supervised developments or large-scale government projects. They were considered part of the royal court, didn't have to fight in the military, and had guaranteed employment not only for themselves, but for their sons as well.[16] In the case of one scribe of the third dynasty's chief, Imhotep, it even meant post-mortem deification.[17]

Later, another era in communication began with the creation of the first form of mass media: Gutenberg's moveable type, fitted to a printing press, which enabled writing to be produced as type-set books, each copy identical to the last, without scribes.

Again, society was transformed. Literacy spread along with printing. As books became plentiful and inexpensive, they could be acquired by any prosperous, educated person, not just by the ruling or religious classes. This set the stage for the Renaissance, the flowering of artistic, scientific, cultural, religious, and social growth that swept across Europe. Next came the revival and spread of democracy. By the time of the American Revolution, printing had made Thomas Paine's pamphlets bestsellers that rallied the troops to victory. The modern metropolitan newspaper, radio, television—all were based on the same basic idea: that communication could be mass-produced from a central source.

The latest transformational change came in earnest just three decades ago, when the personal computer and then the Internet converged to throw us firmly into the digital age. Today, five billion people have cell phones. A constantly flowing electron cloud encircles and unites a networked planet. Anyone with a broadband connection to the Internet has access to much, if not all, of the knowledge that came before, and the ability to communicate not just as a single individual but as a broadcaster. Smartphones are pocket-sized libraries, printing presses, cameras, radios, televisions—all that came before, in the palm of your hand.

The Arguments Against Progress

Technical progress always comes with its critics. The greater the speed and power of this progress, the greater the criticism. Intel researcher Genevieve Bell notes that every time we have shifts in technology, we also have new

---

15 It's important to note, in the context of power's relationship to information, that reading and writing quickly became trade secrets belonging to this set of professionals. Women were quickly excluded. Lower-class citizens needn't apply.

16 Barry J. Kemp. *Ancient Egypt: Anatomy of a Civilization*. Routledge: 2006.

17 M. Lichtheim. *Ancient Egyptian Literature* (p. 104). The University of California Press: 1980, vol. 3.

moral panic. Panic? Here's just one example: there were some who believed, during the development of the railway, that a woman's uterus could go flying out of her body if she accelerated to 50 miles per hour.[18]

Electricity came with a set of critics, too: the electric light could inform miscreants that women and children were home. The lightbulb was a recipe for total social chaos.

These Luddite folk tales are funny, looking back. But other criticisms have gained traction over the centuries.

Our connection to the teachings of Socrates, for instance, is through the written word of Plato, because Socrates was vehemently against the written word. Socrates thought that the book would do terrible things to our memories. We'd keep knowledge in books and not in our heads. And he was right: people don't carry around stories like *The Iliad* in their heads anymore, though it was passed down in a verbal tradition for hundreds of years before the written word. We traded memorization for the ability to learn less about more—for choice.

Critics of the printing press believed that books would cause the spread of sin and eventually destroy the relationship between people and the church. As author and New York University professor Clay Shirky rightfully points out, the printing press did indeed fuel the Protestant Reformation, and yes, growth in erotic fiction.[19]

Though some critiques of the written word have fared better than others, all have faded over time. There just aren't that many people today who think the printing press was a bad idea—not if five billion of them are voting with their purchases to carry one around with them all day.

Despite this, critiques of technology on moral grounds look very similar today. The strongest critiques (as Bell notes) tend to be about women and children. As if it were the modern-day critique of electricity, the television show "To Catch a Predator" features sexual predators using the Internet to seduce children—the subtext is that this powerful new tool can be used to steal your babies.

Still, there is a serious trend emerging in digital age critiques. Distinguished journalists, acclaimed scholars, and prominent activists are worried about what the information explosion is doing to our attention spans or even to our general intelligence. Bill Keller, former executive editor of *The New*

---

18 *http://blogs.wsj.com/tech-europe/2011/07/11/women-and-children-first-technology-and-moral-panic/*

19 *http://www.shirky.com/weblog/2009/03/newspapers-and-thinking-the-unthinkable/*

York Times, equated allowing his daughter to join Facebook to passing her a pipe of crystal meth.[20]

Nicholas Carr's book *The Shallows* (W.W. Norton) is full of concerns that social media is making his brain demand "to be fed the way the Net fed it—and the more it was fed, the hungrier it became."[21] In the *Atlantic*, Carr's "Is Google Making Us Stupid" expresses similar fears: "Over the past few years I've had an uncomfortable sense that someone, or something, has been tinkering with my brain, remapping the neural circuitry, reprogramming the memory."[22]

In his book *The Filter Bubble* (Penguin), my friend and left-of-center activist Eli Pariser warns us of the dangers of personalization. Your Google search results, your Facebook newsfeed, and even your daily news are becoming tailored specifically to you through the magic of advanced technology. The result: an increasingly homogenized view of the world is delivered to you in such a fashion that you're only able to consume what you already agree with.

These kinds of critiques of the Web are nothing new. They're as old as the Web itself—older, actually. In 1995, in the very early days of the World Wide Web, Clifford Stoll wrote in *Silicon Snake Oil* (Anchor), "Computers force us into creating with our minds and prevent us from making things with our hands. They dull the skills we use in everyday life."

Keller, Stoll, and Carr all point to something interesting: new technologies do create anthropological changes in society. Yet some of these critics seem to miss the mark; the Internet is not some kind of meta bogey man that's sneaking into Mr. Carr's room while he sleeps and rewiring his brain, nor did Mr. Stoll's 1995 computer sneak up behind him and handcuff him to a keyboard.

Moreover, the subtext of these theories—and ones like them—is that there may be some sort of corporate conspiracy to try to, as Jobs put it, "dumb us down." Somehow I doubt that Larry Page and Sergey Brin, the founders of Google, woke up one morning with a plan to rewire our brains. Twitter CEO Jack Dorsey is probably not a super-villain looking to destroy the world's attention span with the medium's 140-character limit. Mark Zuckerberg is likely not trying to destroy the world through excessive friendship-building.

---

20 *http://www.nytimes.com/2011/05/22/magazine/the-twitter-trap.html*

21 *http://www.theatlantic.com/magazine/archive/2008/07/is-google-making-us-stupid/6868/*

22 Carr, Nicholas. *The Shallows: What the Internet Is Doing to Our Brains* (p. 16). W.W. Norton & Company: 2010. Kindle Edition.

Blaming a medium or its creators for changing our minds and habits is like blaming food for making us fat. While it's certainly true that all new developments create the need for new warnings—until there was fire, there wasn't a rule to not put your hand in it—conspiracy theories wrongly take free will and choice out of the equation. The boardroom of Kentucky Fried Chicken does not have public health as its top priority, true, but if everyone suddenly stopped buying the chicken, they'd be out of business in a month. Fried chicken left in its bucket will not raise your cholesterol. It does not hop from its bucket and deep-dive into your arteries. Fried chicken (thankfully) isn't autonomous, of course, and isn't capable of such hostility.

As long as good, honest information is out there about what's what, and people have the means to consume it, the most dangerous conspiracy is the unspoken pact between producer and consumer.

Out of the four critiques—those of Keller, Carr, Pariser, and Stoll—Pariser's is the one that makes the most sense to me. Personalization today is mostly a technical issue with consequences that the technologists at our major Internet companies are developing in order to keep us clicking. That said, personalization isn't an evil algorithm telling us what our corporate overlords want us to hear; rather, it's a reflection of our own behavior.

If right-of-center links are not showing up in your Facebook feed, it's likely because you haven't clicked on them when you've had the opportunity. Should corporations building personalization algorithms include mutations to break a reader's filter bubble? Should people be able to "opt out" of tracking systems? Absolutely. But readers should also accept responsibility for their actions and make efforts to consume a responsible, non-homogenous diet, too. The problem isn't the filter bubble, the problem is that people don't know that their actions have opaque consequences.

As with Socrates' reluctance to embrace the written word, critics like Carr and Stoll are onto something, but they're attacking the wrong thing. It wasn't the written word that has stopped most of us from memorizing the epic Odyssey; rather, it is our *choice* not to memorize it anymore, and to read books instead.

Anthropomorphized computers and information technology cannot take responsibility for anything. The responsibility for healthy consumption lies with human technology, in the software of the mind. It must be shared between the content provider and the consumer, the people involved.

There Is No Such Thing as Information Overload

Once we begin to accept that information technology is neutral and cannot possibly rewire our brains without our consent or cooperation, something else becomes really clear: there's no such thing as *information overload*.

It's the best "first world problem" there is. "Oh, my inbox is so full," or, "I just can't keep up with all the tweets and status updates and emails" are common utterances of the digital elite. Though we constantly complain of it—of all the news, and emails, and status updates, and tweets, and the television shows that we feel compelled to watch—the truth is that information is not requiring you to consume it. It can't: information is no more autonomous than fried chicken, and it has no ability to force you to do anything as long as you are aware of how it affects you. There has always been more human knowledge and experience than any one human could absorb. It's not the total amount of information, but your information habit that is pushing you to whatever extreme you find uncomfortable.

Even so, we not only blame the information for our problems, we're arrogant about it. More disturbing than our personification of information is the presumption that the concept of information overload is a new one, specific to our time.

In 1755, French philosopher Denis Diderot noted:

> *"As long as the centuries continue to unfold, the number of books will grow continually, and one can predict that a time will come when it will be almost as difficult to learn anything from books as from the direct study of the whole universe. It will be almost as convenient to search for some bit of truth concealed in nature as it will be to find it hidden away in an immense multitude of bound volumes."*[23]

Diderot was on target with the continuous growth of books, but he also made a common mistake in predicting the future. He presumed that technology would stay complacent. In this short verse, he didn't anticipate that with an increasing number of books, new ways to classify and organize them would arise.

A century after Diderot wrote, we had the Dewey Decimal system to help us search for those bits of truth "hidden away in an immense multitude of bound volumes." Two and a half centuries later, the pages are bound not to

---

23 *http://books.google.com/books?id=z5MeICA-zzIC&pg=PA85&lpg=PA85&dq=As+long+as+the+c enturies+continue+to+unfold,+the+number+of+#v=onepage&q=As%20long%20as%20the%20 centuries%20continue%20to%20unfold%2C%20the%20number%20of&f=false*

bookbindings, but to electronic formats. It has never been faster and easier than with Amazon to find and buy a book in either a print or electronic version. And Google would be delighted if every word of every book were searchable—on Google.

To say, therefore, that the Internet causes our misinformation ignores history. In the modern arms race between fact and fiction, it's always been a close fight: we're no better at being stupid or misinformed than our grandparents were. It's the ultimate ironic form of generational narcissism. History is filled with entire cultures ending up misinformed and misled by ill-willed politicians and deluded masses.

Just like Stoll and Carr, Diderot was onto something, but he was lured into the trap of blaming the information technology itself.

The field of health rarely has this problem: one never says that a lung cancer victim dies of "cigarette overload" unless a cigarette truck falls on him. Why, then, do we blame the information for our ills? Our early nutritionist, Banting, provides some prescient advice. He writes in *Corpulence*:

> "*I am thoroughly convinced, that it is quality alone which requires notice, and not quantity. This has been emphatically denied by some writers in the public papers, but I can confidently assert, upon the indisputable evidence of many of my correspondents, as well as my own, that they are mistaken.*"[24]

Banting's letter gives us an idea of what the real problem is. It's not information overload, it's information *overconsumption* that's the problem. Information overload means somehow managing the intake of vast quantities of information in new and more efficient ways. Information overconsumption means we need to find new ways to be *selective* about our intake. It is very difficult, for example, to overconsume vegetables.

In addition, the information overload community tends to rely on technical filters—the equivalent of trying to lose weight by rearranging the shelves in your refrigerator. Tools tend to amplify existing behavior. The mistaken concept of information overload distracts us from paying attention to behavioral changes.

The Information Overload Research Group, a consortium of "researchers, practitioners and technologists," is a group set up to help "reduce information overload." Its website offers a research section with 26 research papers on the topic, primarily focused on dealing with electronic mail and technology used to manage distractions and interruptions. If they

---

24 *http://www.lowcarb.ca/corpulence/corpulence_full.html*

mention user behavior at all, they're focused on a person's relationship with a computer and the tools within it.

Now, don't get me wrong. I appreciate a good spam filter as much as the next person, but what we need are new ways of thinking and of coping.

Just as Banting triggered a wave of concern about diet as we shifted from a land of food scarcity to abundance, we have to start taking responsibility ourselves for the information that we consume. That means taking a hard look at how our information is being supplied, how it affects us, and what we can do to reduce its negative effects and enhance its positive ones.

CHAPTER 3

Big Info

"For 200 years the newspaper front page dominated public thinking. In the last 20 years that picture has changed. Today television news is watched more often than people read newspapers, than people listen to radio, than people read or gather any other form of communication. The reason: people are lazy. With television you just sit-watch-listen. The thinking is done for you."

—Anonymous memo, Nixon Presidential Archives
Largely attributed to Roger Ailes,
Nixon Campaign Staffer and now FOX News Chairman[25]

The year 1960 was the year that television became the most important thing in politics. After refusing to wear makeup and campaigning for hours beforehand, Richard Nixon appeared weary, sick, and sloppy next to the well-rested and confident John Kennedy. Seventy million people tuned into the first televised presidential debate, and after it was over, John Kennedy moved into the lead and never looked back.

Having learned his lesson, when he ran for president again in 1968, Nixon hired a 28-year-old local television producer from Cleveland to be the media advisor to his campaign. His name was Roger Ailes, and he'd take Richard Nixon from the sickly sideliner to the polished, professional candidate who made it to the White House.

We have to put this into context a bit: there weren't two generations of people in America who grew up with televisions in the household like there are today. Television for many was as magical and mysterious as the Internet is now. It was a new frontier, and like the social media consultant's relationship with Washington today, there was a rising class of consultants

---

25 http://gawker.com/5814150/roger-ailes-secret-nixon+era-blueprint-for-fox-news

preaching the gospel of the new medium to candidates and politicians eager to get in on the action.

After the success of Nixon's '68 campaign, Ailes quickly rose to power inside and outside the White House. He launched Roger Ailes Associates to help right-of-center candidates get elected, and advised the president on media and political strategy. For Nixon, he did everything from directing the television broadcast of the White House Christmas Tree Lighting Ceremony to suggesting that the administration infiltrate the George Wallace campaign in order to "guard their flank."[26]

It was during the Nixon administration that Ailes had the idea for a "pro-administration news system," recognizing that Washington was close to three major airports (Dulles, National, and Baltimore) and that video footage could get to any major media market in the country. Ailes believed that the media had become dominated by negativity and said that the failure of business leaders to translate their agenda into something that ordinary Americans could understand was responsible for a cancer that was killing America. It was the blueprint for what eventually became Fox News.

Today, if you ask the Democratic party establishment in Washington whom they hate the most, you'll likely find Roger Ailes near the top of the list. In 1996, he was tapped by the News Corp chairman to launch and manage Fox News, now the number one cable news channel in the country. According to the *New York Times*,[27] Ailes' network makes more money than CNN and MSNBC, plus the nightly news broadcasts of the major networks, combined.

In under a decade, Ailes quickly toppled the other news broadcasts with less money, fewer reporters, and far less infrastructure than anyone else. So how'd he do it?

Ailes knew two things that nobody else did: first, that cable news was different than broadcast news. Because there were more channels available on cable than broadcast, and because of the nature of the medium, you didn't have to try to please all your viewers; you could pick and grow a niche audience. Second, like many other conservatives today, Ailes felt the media was eliminating his point of view; Ailes knew that cable could provide an alternative news source.

He couldn't compete with CNN on news. He had to compete with them by providing a different choice altogether. Ailes built himself a media network

---

26 *http://gawkernet.com/ailesfiles/ailes2.html*

27 *http://www.nytimes.com/2010/01/10/business/media/10ailes.html*

PART I: INTRODUCTION

that, in his mind, didn't eliminate the conservative point of view, and he found that much of America wanted it. Ailes found himself in a perfect spot: building a super-profitable business that aligned itself with his values. For News Corp Chairman Rupert Murdoch, it was even better. Ailes figured out how to build a hugely profitable cable news outlet without having to pay for the infrastructure of a CNN. Giving people what they want is far more profitable than giving them the facts. In his own words:

> "I can't look hip, but I don't want to be hip. And yet, you talk about programming a channel—I could out-program these thirty-year-olds in terms of what needs to be on there, how to get to the audience, how to get to younger people. I speak at colleges. Whatever it is, I always tell them, look, I can out-program you. I'll challenge them all the time."[28]

Now, presiding over the Fox News empire, Ailes doesn't want to make news—that's not what he's good at. What he wants to do is give people what they want: entertainment and affirmation.

The Pew Center for Excellence in Journalism[29] estimates that Fox News spends 72% of its budget on program expenses (expenses tied to specific programs, like host salaries) and 27.8% of its expenses on administrative expenses (things like newsrooms). CNN, on the other hand, spends 56% of its expenses in the administrative category, and 43.9% on program expenses. CNN has a total staff of 4,000 people working in its studios and 47 bureaus. Fox News has 1,272 members of staff in just 17 bureaus.

The strategy is simple: it's cheaper to pay one media personality a two million dollar salary than it is to pay 100 journalists and analysts $40,000 a year. What's better, people like hearing their beliefs confirmed more than they like hearing the facts. For Murdoch and Ailes, it must have been like discovering the McDonald's business model. People like french fries more *and* they're cheaper to make than steamed broccoli! That's sound business.

Ailes hasn't just made Fox News a media empire; he's changed the news industry. In order to stay competitive, the other cable networks and news services have had to change their strategies. In 2005, MSNBC started to see the dollar signs and began investing in programming costs over newsroom costs. In 2005, it spent 58% of its costs on programming expenses, but by 2010 that number sat at 88%. It brought in personalities like Chris Matthews, Joe Scarborough, Keith Olbermann, and Rachel Maddow. The results are astounding: MSNBC surpassed CNN's viewership in 2010

---

28 *http://www.esquire.com/blogs/politics/roger-ailes-quotes-5072437#ixzz1VJICt1q4*
29 *http://stateofthemedia.org/2011/cable-essay/*

with a staff of just 600 people in 4 bureaus.[30] Ailes' experiment didn't just succeed for Fox News. It has been proven correct for MSNBC too.

CNN, on the other hand, took a different path. Watching the other two networks go their right- and left-of-center ways, CNN figured there must be some room left for the facts.

Over the same period of time, CNN canceled much of their hard personality-driven shows—like "Crossfire," starring Democratic party media consultants Paul Begala and James Carville versus right-leaning pundits Tucker Carlson and Robert Novak—and replaced them with less overtly opinionated anchors like John King, Anderson Cooper, and Wolf Blitzer.

While the network still leverages its vast bureaus to achieve high ratings in times of international conflict, like 2010's Arab Spring, the result was an astounding plummet in the ratings. In prime time, CNN is now a third-place network—it's beaten by MSNBC and Fox's personality-driven journalism night after night. Fox News is in first place on the right, and MSNBC is second on the left. CNN sits at the bottom in the middle, providing real news that nobody wants to hear.[31]

Although CurrentTV has been around for years now, it too has recently jumped on the bandwagon. After MSNBC and anchor Keith Olbermann had a fairly public dispute, Current snapped up the partisan media personality, and has seen an increase in market share, and an increase in profit.

Our news networks have turned into affirmation distributors.

Gone are the days of Edward R. Murrow—of journalists seeking to deliver truth to their audiences. Instead, the advertisers and their salesforces have taken over. The sales teams want to sell you the delicious stuff that you keep coming back for more, even if it's at the expense of the truth.

Like our food companies, our media companies—the companies that produce and deliver much of the information that we consume—have been consolidated down to a handful: Time Warner, AOL, Disney, Viacom, News Corporation, CBS Corporation, and Comcast. Together, these companies represent many of the movies we watch, newspapers we read, magazines we subscribe to, books we buy, and Internet services we use.

---

30 This analysis comes from the excellent Pew Research Center's Project for Excellence in Journalism State of the News Media reports, 2004–2001. *http://stateofthemedia.org/previous-reports/*

31 *http://www.mediabistro.com/tvnewser/category/ratings*

These companies are all publicly traded corporations—ones with responsibilities to their shareholders to do their best to maximize profits. It's called *fiduciary responsibility*, and this means driving down costs, increasing revenues, and growing the company. Every board member and every officer of a large corporation must grow their company and maximize shareholder wealth or face unemployment.

Food companies want to provide you with the most profitable food possible that will keep you eating it—and the result is our supermarket aisles filled with unimaginable ways to construct and consume corn. Media companies want to provide you with the most profitable information possible that will keep you tuned in, and the result is airwaves filled with fear and affirmation. Those are the things that keep the institutional shareholders that own these firms happy.

Choice Lessons

These issues aren't new. In 1790, Benjamin Franklin's grandson, Benjamin Franklin Bache, found himself the proud inheritor of all of his grandfather's printing equipment and books. He quickly set up the Philadelphia Aurora, stating that "This paper will always be open, for the discussion of political, or any other interesting subjects, to such as deliver their sentiments with temper and decency, and whose motives appears to be, the public good."[32] Or, like the now familiar Fox News slogan, that it would be "Fair and Balanced."

Over the course of the next decade, Bache used his paper to denounce President Adams' administration, and Adams' party: the Federalists. The Federalists passed the Sedition Act in 1798, which made it a crime to publish "false, scandalous, and malicious writing," and quickly arrested Bache. Whether Bache's accusations were true or not didn't matter: public outrage ensued, and in the election of 1800, the Federalists didn't just lose—their party was all but destroyed.

What's different today is that new tyranny of the majority is more efficient than it used to be. It's driven in real time by the tiny but meaningful transactions we have with our media providers every day. That's why the world of politics is dividing into the world of MSNBC and DailyKos versus Fox and Andrew Breitbart. We now have the option to participate in the news realities we want to tune into, with the tribes we elect to be part of.

---

32 Smith, Jeffery A. Franklin and Bache: Envisioning the Enlightened Republic (p. 102). Oxford University Press: 1990.

The New Media

Even more than television, Fox routinely tweaks the news on the Web to make the news more palatable to its audience. Even when it takes content from other sources like the Associated Press and puts it on its website, the organization tweaks the headlines to make them more attractive to its conservative audience. The AP's story "Economic Worries Pose New Snags for Obama" turned into "Obama Has a Big Problem with White Women." "Obama to Talk Economy, Not Politics, in Iowa" turned into "White House Insists Obama's Iowa Stop for Economy, Not 2012." And "Malaysia Police Slammed for Cattle-Branding Women" turned into "Malaysian Muslims Cattle-Brand Prostitutes."

Fox isn't about advancing a conservative agenda. For its parent, News Corporation, it's about the dollars. Fox changes these headlines on the Web not because it has an agenda, but because people click on them more, meaning that more advertisements can be shown, and more money can be made. And Fox's headline tweaking is just the beginning. With the Web, our choices aren't even bound by the number of channels our cable boxes offer. With the Web, our choices are limitless.

Of course Foxnews.com isn't the only web operation that does this. The *Huffington Post* is also into these shenanigans. On any given day, the *Huffington Post*'s homepage is a bizarre sight: a defense of *New York Times* op-ed columnist Paul Krugman coupled with the "Top Embarrassing Photos of Obama's Vacation." "A Computer Chip Mimics the Human Brain," it tells me, next to the warning: "Don't Go Shopping with People Harder Than You." Along the sidebar, we're treated to images of celebrity wardrobe malfunctions and "make out sessions."

These things are there, not because of Arianna Huffington's contempt for the public, but because we click on them, and we click on them more than we click on anything else. The *Huffington Post* is a reflection of its readership's interests. In just writing this bit about the site, I've found myself lost in its enormous sea of link-bait. There's so much I need to know that I didn't know I needed to know!

The *Huffington Post* has turned content-creation on its head, using technology to figure out what it is that people want, and finding the fastest way to give it to them. Just like the Cheesecake Factory tests its delicious cheesecakes in a test lab to make sure they're delicious before they are set in front of you, the *Huffington Post* uses your behavior to understand what you want. Unlike the Cheesecake Factory though, they can do it in real-time.

They employ a technology called *multivariate testing* (or *A/B testing*) to figure out what users want in near real time. According to Paul Berry, CTO, the site randomly displays one of two headlines for the same story for five minutes. After the elapsed time, the version with the most clicks wins and everybody sees that one. The result is the same: sensational headlines. "Wisconsin Protests Have State GOP Sending State Troopers After Democrat" turned into "GOP Sends National Guard After Dem Leader."[33]

The Huffington Post's parent company, America Online, is far from its dial-up and busy signal roots. AOL makes its money by acquiring content and selling advertisements. In 2011, it ranked as the fifth largest property online in the United States behind Google, Yahoo, Microsoft, and Facebook. It reaches, in any given month, over a third of the United States population: about 110 million people.

The New Journalists

The industrialization of information is doing to journalists what the industrialization of farming did to farmers. In an effort to squeeze every bit of profit out of a piece of content, expensive journalists are being replaced by networks of less-qualified but much cheaper independent contractors. In the world of fiduciary responsibility, quality journalism means market inefficiency.

Though it still makes money from its Internet service provider business, today AOL is what's known online as a *content farm*, and it shares a lot in common with its agricultural counterpart, the factory farm. AOL's content is driven by a policy known as "The AOL Way," a document in the form of a Powerpoint presentation that was leaked from AOL in early 2011. The AOL Way instructs the entire content arm of AOL on how it should operate.

The intent of The AOL Way is to decrease the costs and increase the profitability of the content the company produces. According to the plan, each editor should use four factors to decide what to cover: traffic potential, revenue potential, turn-around time, and at the bottom of the list, editorial quality. All editorial content staff are expected to write 5 to 10 stories per day, each with an average cost of $84, and a gross margin (from advertising) of 50%.

In short, it's the job of the writer to produce popular content as cheaply and quickly as possible. That explains why the front page of AOL.com features the headline "Watch: Orangutan Gets Even With Rude Lady"; asks me to

---

33 *http://www.theblaze.com/blog/2011/02/18/really-huffington-post-really/*

guess the age of the world's oldest female bodybuilder; and offers me "Ten Bizarre Mosquito Prevention Tips."

At the heart of The AOL Way is a technology called BlogSmith. It's a software platform that allows editors to generate and produce content and measure their impact on the revenue and profitability of the network. AOL's editors are instructed[34] to first use BlogSmith's Demand module to identify topics in demand. BlogSmith looks at search query volume and breaks terms up into three categories: *breaking* (current trending topics), *seasonal* (topics historically in demand during certain time periods), and *evergreen*—topics that are consistently in demand across all AOL products.

Editors are then assigned these categories by their managers, and instructed to quickly write content matching these topics. (If management expects 5 to 10 posts per working day, then that's about one post per hour.) Each post is to be tagged with popular search terms so that they're more easily discoverable by search engines. Sarah Palin's ride through downtown D.C. on Memorial Day was tagged on AOL-owned *Huffington Post* as: "2012 Election, Sarah Palin 2012, Elections 2012, Sarah Palin, Sarah Palin For President, Palin 2012, Palin Bus Tour, Palin For President, Palin Motorcycle, Rolling Thunder, Sarah Palin Bus Tour, Sarah Palin Motorcycle, Politics News" to cover all the search bases.

BlogSmith then carefully tracks the return on investment. Under its *performance* tab, it tells the author that it cost $15 to make the piece of content, and it's returned $82.95 in advertising. In big green letters it tells the editor they've made $67.95 in profit for the mothership. It's journalism, commoditized.

So, why this setup? Here's what one AOL writer—John Biggs—had to say on AOL blog TechCrunch.com:

> *"There's no money in shaking the crown of power from a lowly perch. There is money in feeding novel info to a ravenous, neophilic audience."*[35]

They do it because it works! The headlines are irresistible. In doing the research for this chapter alone, I've watched a one-and-a-half-minute short film on Lindsay Lohan, seen John Lithgow's dramatic interpretation of a press release from Newt Gingrich's presidential campaign, learned that Shiloh Jolie-Pitt turned five years old, and yes, have seen a lot of pictures

---

34 *http://www.businessinsider.com/the-aol-way#-14*

35 *http://techcrunch.com/2011/08/17/journalist-crowdsources-an-article-about-a-crowdsourcing-company-hilarity-ensues/*

and videos of Sarah Palin riding a motorcycle. The age of the oldest female bodybuilder as of this writing, by the way, is 74.

These articles aren't written by people with a journalism background. They're written by freelancers—independent contractors—who needn't be provided any healthcare or retirement benefits. For content farmers, they're simply credited—about $15 for a written piece of content, $20 for a video—directly to their bank account. Copy editors are paid a remarkable $2.50 per piece of content. Traditional newspapers pay about $300 to a freelance journalist for the same amount of work.

The jobs themselves tend to be no piece of cake. According to former AOL employee Oliver Miller:

> "My 'ideal' turn-around time to produce a column started at thirty-five minutes, then was gradually reduced to half an hour, then twenty-five minutes. Twenty-five minutes to research and write about a show I had never seen—and this twenty-five minute period included time for formatting the article in the AOL blogging system, and choosing and editing a photograph for the article. Errors were inevitably the result. But errors didn't matter; or rather, they didn't matter for my bosses.
>
> I had panic attacks; we all did. My fellow writers would fall asleep, and then wake up in cold sweats. I worked the graveyard shift—11PM to 7 or 8AM or later—but even the AOL slaves who wrote during the day would report the same universal experience. Finally falling asleep after work, they would awake with a jump, certain that they had forgotten something—certain that they hadn't produced their allotted number of articles every thirty minutes. One night, I awoke out of a dead sleep, and jumped to my computer, and instantly began typing up an article about David Letterman. I kept going for ten minutes, until I realized I had dreamed it all. There was no article to write; I was simply typing up the same meaningless phrases that we all always used: 'LADY GAGA PANTLESS ON LATE NIGHT WITH DAVID LETTERMAN,' or some such.
>
> Then there was the week where I only slept for about six hours over the course of five days—a week that ended with me being so exhausted that I started having auditory hallucinations, constantly hearing a distant ringing phone that didn't exist, or an imaginary door slamming in the background."[36]

---

36 *http://thefastertimes.com/news/2011/06/16/aol-hell-an-aol-content-slave-speaks-out/*

Now granted, Miller was still working from the comfort of his house. You can't compare the job to farming at the physical level—as we've noted before, farming tends to be one of the more dangerous professions in America. Mr. Miller isn't going to lose his life in a tragic blogging accident. But it also doesn't sound like it was a particularly nice job to have.

While factory farming dominates agriculture, content farming now dominates our information consumption online. Its industrialization goes far beyond news.

If you've ever searched online for how to change the oil in your car, how to iron a shirt, or how to unclog a toilet, chances are you've run across a website called eHow.

eHow is owned by another content farm called Demand Media—probably the largest, in terms of workforce, of all content farms. They supply the content to eHow, Lance Armstrong's *LiveStrong.com*, and Tyra Banks' *typeF.com*. Beyond their own sites, Demand Media also provides farmed content to a variety of websites across the Internet. In terms of traffic, Demand Media's sites receive more unique visits than Fox News's online presence and the *Washington Post* combined. It's the 18th largest property on the Internet. Nearly four million more people online visit a Demand Media website than visit Craigslist in a given month.[37]

Content farms are big businesses. As of this writing, AOL is worth $1.2 billion. Demand Media is worth $663 million. Associated Content—the content farm once billed "The People's Media Company"—sits as part of *Yahoo.com*.

Seek and We Shall Profit

Search traffic is the fuel for most of these operations. Content farms have solved a fairly simple math problem. Google and other search engines provide data about the top searches for any given moment. Content farms write articles that then show up in the top results for those searches, and show ads on the pages that those articles are on. Over a short period of time, those pieces of content show enough ads so that they're profitable. It may cost $15 to make a piece of content, but if every view of the page that it's on earns you a nickel in advertising revenue, you only need 301 views to start turning a profit.

---

37 *http://www.comscore.com/Press_Events/Press_Releases/2011/9/comScore_Media_Metrix_Ranks_Top_50_U.S._Web_Properties_for_August_2011*

Google is both an accomplice and a benefactor of content farms. On one hand, Google is a search company. It has a vested interest in making its search experience high quality: if you search for how to change the oil in a 1976 Chevy Nova, Google wants you to get good, clear advice as a result of your search. On the other hand, Google is an advertising company. As of 2011, Google controlled 43.5% of total online advertising spending. Lining the sides of sites like Demand Media are advertisements provided by Google's ad network: Google's getting a cut of the site's advertising revenues, to the tune of millions of dollars.

For now, Google is opting to take the high road and the long-term view. It's in Google's interest to give great results to its users—having the Web littered with poor-quality information sources ends up making Google itself less relevant. Thus Google is taking steps to increase incentives to reduce the farming sites' influence on Google search results.

In early 2011, it released new search technology code-named Panda to curb the effectiveness of content farms. The effect: 17% of Demand Media's keywords were dropped from the first page of Google's search results. The *New York Times* dubbed it "Google's war on nonsense."

Panda's release and deployment is just a battle, not the end of this war. With the money at stake, this is an arms race, not a problem that gets completely solved with some algorithm changes. As Google fine-tunes its algorithm, companies like AOL and Demand Media will find new ways to achieve higher rankings on results pages.

There's also an ethics problem Google must answer. Google can't overstep their bounds in the war on nonsense in deciding what is nonsense and what is not. Should Google start providing too much editorial guidance on its top ten results, for instance, it could run into political problems: a search for "climate change" that isn't seen as fair by either side of that debate may trigger a conversation about regulation. Around Washington there are already whispers of "search neutrality" on this very subject.[38]

There's another big problem outside of Google's reach, an untapped and potentially huge market for these content farmers: your local, printed newspaper.

If a traditional freelancer for USA Today makes $300 for the paper, why not take that freelancer and replace him with a content farmer? If you're a newspaper editor, you may soon be facing that choice. Facing plummeting advertising sales, and the complete gutting of classified ads, traditional

---

38 *http://www.searchneutrality.org/*

media is taking notice: *USA Today*'s TravelTips, the *San Francisco Chronicle*'s Home Guides, and the *Houston Chronicle*'s "Small Business Resource Center" are now powered by the Demand Media content farm. It may be only a matter of time before these articles start making their way off the Web and into your paper, or onto the 11 o'clock news.

Churnalism

According to the Bureau of Labor Statistics at the Department of Labor, in 2008 there were just 69,300 news analysts, correspondents, and reporters in the United States, earning a median average wage of $34,850.[39] The department expects a 4% decline in the sector between 2008 and 2018. The journalist is going the way of the farmer.

The world of public relations is moving in the opposite direction. In 2008, there were over 275,200 public relations specialists in the United States, earning a median annual wage of $51,280. The Bureau expects the field to continue to grow, upwards of 25% between 2008 and 2018.

As a result, our reporters are suffering from information obesity themselves. For every reporter in the United States, there are more than four public relations specialists working hard to get them to write what their bosses want them to say. That's double what there were in 1970. Journalists are assaulted with press releases stuffed in their mailboxes, polluting their email inboxes, and pouring out of their fax machines, full of pitches, sound bites, and spin.

In an effort to cut costs, journalists often become more filters than reporters, succumbing to the torrents of spin heading their way, and passing on what's said by the scores of PR consultants. Rather than report the news, they simply copy what's in a press release and paste it into their stories. It's a kind of commercially advantageous and permissible plagiarism called *churnalism*.

In 2009, the independent filmmaker Chris Atkins decided to put this phenomenon to the test. He worked with a small web design firm to set up a fake website for a PR agency, and a fake male cosmetic product, the Penazzle, sold by a fake company: MaleBeautyDirect.com. It's a temporary tattoo of sorts that sits on a male's waxed lower abdomen.

Atkins sent press releases out to all the large newsdesks of large newspapers in the UK, and had researchers follow up with calls to each targeted member of the press. The next day, the Sun, the largest newspaper in Britain, carried

---

39 *http://www.bls.gov/oco/ocos088.htm*

the story with the headline "Nobbies Dazzlers" with 45% of the text lifted straight from the press release. He demonstrated the same thing repeatedly, with a story of a fake "chastity garter" that secretly texts a boyfriend when she's cheating on him, and a fake story of how the British Prime Minister's cat Larry had been stolen from its rightful owner. In every case, the story was carried, without fact checking, and largely copied and pasted into the press.

It's a widespread problem—and not just the problem of hoaxters either. Researchers at Great Britain's Cardiff University found that upwards of 60% of press articles and 34% of broadcast stories were the results of churnalism.

A website has even been set up, *churnalism.com*, that allows you to paste in the copy of news articles to see how an article has been "churnalized"; how many times a press release has been copied and pasted across various sources in the UK.

The parallels between how our media has changed and how agriculture changed are obvious if you look closely: what happened to farmers is happening to journalists. What happened to our diets is happening to our news. And like with our food, there's not much we can do about it; the draw of living with abundant supply is too strong, and too beneficial, to fight. Instead, we've got to understand how to cope in a world with different rules.

CHAPTER 4

We Are What We Seek

"A fanatic is one who can't change his mind
and won't change the subject."

—Winston Churchill (attributed)

I hate to bring up partisan politics; it generally doesn't do much good when one is trying to make any form of argument. But it turns out that if I look back at the times in my life when I have had a recognizably bad information diet, they're the times when I've been knee-deep in politics. Working in politics is an amazing opportunity to try to affect change, sure. But it's also a great way to pick up a disease called *delusion*.

In the summer of 2003, I packed my bags and headed up to New England to work as the lead programmer for the insurgent presidential candidate, Howard Dean. The staff was reasonably kind—mostly native Vermonters and interns at the time. They liked to pick on this poor Southerner, though; at one point, someone warned me that if I spent too much time outside with my eyes open in the winter, the fluid in my eyeballs would freeze over. I remember shutting my eyes hard and sprinting out across the ice to my car and grasping for the door handle blindly on several occasions. Yankees are tricky, I tell you what.

Cults, startups, Apple keynotes, and political campaigns all have one thing in common: a group of people with delusional loyalty to the mission they're trying to accomplish. Those of us on the Dean campaign feasted on a diet consisting of the narrative that we would be the ones to remove the evil George W. Bush from office. I ended up gaining a lot from that campaign: about 32 pounds from the constant supply of campaign-contributed Ben and Jerry's ice cream, and a healthy dose of crazy.

Each morning, the *media miners*—the folks in charge of watching all the cable news—would feed us clips that told us how well we were doing. The afternoon was filled with blog posts from across the Internet talking about how revolutionary our campaign was. Evenings were filled with watching the latest and greatest episodes of "The West Wing" starring President Bartlett—the fictional president that we assured ourselves was based on Howard Dean, despite producer and writer Aaron Sorkin donating twice as much money in 2004 to the presidential campaigns of Dick Gephardt, Wesley Clark, and John Edwards.

There was also constant speculation: Republican Strategist Karl Rove had said, gleefully, that Dean was the candidate he wanted to win the Democratic nomination. We were emboldened by his claim. They were afraid of us—*Karl Rove never says what he means. He must be giving us his endorsement because he doesn't want to face us!* We'd try to find as many facts as we could to support this idea.

That CNN cut to Donald Rumsfeld instead of showing Howard Dean's speech on tax policy? Certainly evidence that the White House was using whatever it could to keep us off the air. Obviously CNN, too, had become an instrument of this evil republican regime.

The week before the Iowa caucuses, I remember asking the campaign's pollster, Paul Ford, by how much we were going to win Iowa. His response was: "We're not. John Kerry is going to win it by 18 points."

My jaw dropped. I wasn't sad or disappointed. I was mad at Paul and a little disappointed in him. How could he be such a traitor? Hadn't he seen the news? He clearly was incompetent. Any fool could see that we'd correctly leveraged the Internet in Iowa and this puppy was in the bag. Howard Dean would win Iowa and go on to beat George W. Bush.

But Paul was right and we were crazy. You know the rest of the story: Howard Dean lost the Iowa caucus by nearly 20 points, and would go on to give a concession speech with a yell that became his defining moment. Only the political intelligentsia would remember his use of the Web. The rest of the electorate remembers him for that terrible scream.

The morning after the caucuses, our Burlington, Vermont, offices were filled with more delusion. One of my colleagues ran up to me as I walked into the office and said, "Clay, did you see the Governor's speech last night? It was awesome. He's totally back. We're going to win this thing."

We redoubled our efforts—though Dean was down by double-digits in New Hampshire, we could make a comeback. Every primary and caucus after that, we convinced ourselves we still had it. As the weeks went by, as the sinking feeling got stronger that we would lose to John Kerry, we got hungrier and hungrier for any poll that would give us even a slim chance of winning.

If, a month later, you had polled the staff to ask who would win the Wisconsin primary—our line in the sand—we'd have told you it was

Howard Dean. And we'd believe, out of desperation, anything that told us we were right.

We came in third.

Reality Dysmorphia

In eating disorder treatment centers, a physician will often ask the patient to draw an outline of her own body on a large chalkboard. Then, the doctor will ask the patient to place her back against the wall, and trace the actual outline of her body. For many patients, the outlines that they draw are quite exaggerated, sometimes twice as large as their actual bodies.

It's a phenomenon called *body dysmorphia*: that someone's self-image isn't attached to reality. The phenomenon goes beyond the patients just thinking they're a different shape than they really are, though: when the victims of this disorder look in the mirror, they're literally seeing something different than what everybody else does.

During the Dean campaign, the delusion that resulted from my poor information diet was a cognitive version of this disease: reality dysmorphia. I haven't met a single campaign operative here in Washington, D.C., on either side, that didn't have at least a mild case of it.

This kind of delusion comes from psychological phenomena like *heuristics, confirmation bias,* and *cognitive dissonance.*

It turns out our brains are remarkable energy consumers. Though it typically represents only 2% of the human body's weight, the brain consumes about 20% of the body's energy resources.[40] As such, we've evolved—both for our brain's energy consumption, and for our social survival—to use shortcuts in order to be able to handle more complex thoughts.

Think of a *heuristic* as a rule of thumb: a mental shortcut, or the thing you get once you burn your hand on a hot pan and learn that you shouldn't touch hot pans anymore. You needn't bother testing this hypothesis anymore; you know it. Heuristics are psychologically there so that you don't have to think about them anymore, and you can spend your brain's energy thinking about something else.

Heuristics have a dark side, though: they cause us to have unconscious biases towards things we're familiar with, and choose to do the same thing we've always done rather than do something new that may be more efficient.

---

40 Clark, DD; Sokoloff L (1999). Siegel GJ, Agranoff BW, Albers RW, Fisher SK, Uhler MD. ed. *Basic Neurochemistry: Molecular, Cellular and Medical Aspects.* Philadelphia: Lippincott. pp. 637–70. ISBN 9780397518203.

They cause us to make logical leaps that take us to false conclusions. For instance, these mental shortcuts underpin our capacity for racism, sexism, and other forms of discrimination.

One such nefarious heuristic is called *confirmation bias*. It's the psychological hypothesis that once we begin to believe something, we unconsciously begin seeking out information to reinforce that belief, often in the absence of facts. In fact, our biases can grow to be so strong that facts to the contrary will actually strengthen our wrong beliefs.

In 2005, Emory University professor Drew Westen and his colleagues recruited 15 self-described strong Democrats and 15 strong Republicans for a sophisticated test. They used a functional magnetic resonance imaging (fMRI) machine to study how partisan voters reacted to negative remarks about their party or candidate. Westen and his colleagues found that when these subjects processed "emotionally threatening information" about their preferred candidates, the parts of the brain associated with reasoning shut down and the parts responsible for emotions flared up.[41] Westen's research indicates that once we grow biased enough, we lose our capacity to change our minds.

Following Westen's study, social scientists Brendan Nyhan and Jason Reifle conducted a new test,[42] and discovered what they believe is a "backfire effect."

Nyhan and Reifle provided the subjects with sample articles claiming that President Bush stated that tax cuts would create such economic growth that it would increase government revenues. The same articles included corrective statements from a 2003 Economic Report of the President and various other official sources, claiming that this was implausible. The researchers then showed the students the actual tax revenues as a proportion of GDP *declining* after Bush's tax cuts were enacted.

The results were fascinating: after reading the article, the conservatives in the study were still more inclined to believe that tax cuts *increase revenue* as a result of reading the correction. Hearing the truth made conservatives more likely to agree with the misperception. The facts backfired.

We already know that things like confirmation bias make us seek out information that we agree with. But it's also the case that once we're entrenched in a belief, the facts will not change our minds.

---

41 *http://www.psychsystems.net/lab/06_Westen_fmri.pdf*

42 *http://www-personal.umich.edu/~bnyhan/nyhan-reifler.pdf*

Politics is the area in which scientists have studied the psychological causes of bias the most. It's easy to get people to self-identify, and universities tend to have more of an interest in political science than in other realms of social studies. But you can also see the results of this kind of bias in areas other than politics: talk to a Red Sox fan about whether or not the Yankees are the best team in baseball's history, and you'll see strong bias come out. Talk to MacBook owners about the latest version of Windows and you may see the same phenomenon.

We've likely evolved this way because it's safer. Forming a heuristic means survival: watching your caveman friend eat some berries and die doesn't make you want to conduct a test to see if those berries kill people. It makes you want to not eat those berries anymore, and to tell your friends not to eat those berries either.

Cognitive scientists Hugo Mercier and Dan Sperber took reasoning and turned it on its head. After all, if all the evidence around reasoning shows that we're actually pretty bad at using it to make better choices, then maybe that's not reason's primary function. In their paper "Why do humans reason?,"[43] they argue instead that "reasoning does exactly what can be expected of an argumentative device: Look for arguments that support a given conclusion, and, ceteris paribus, favor conclusions for which arguments can be found." Mercier and Sperber argue that our minds may have evolved to value persuasion over truth. It certainly is plausible— human beings are social animals, and persuasion is a form of social power.

The seeds of opinion can be dangerous things. Once we begin to be persuaded of something, we not only seek out confirmation for that thing, but we also refute fact even in the face of incontrovertible evidence. With confirmation bias and Nyhan and Reifle's backfire effect in full force, we find ourselves both addicted to more information and vulnerable to misinformation for the sake of our egos.

This MSNBC Is Going Straight to My Amygdala

Neuroscience is the new outer space. It's a vacuum of promise and fantasy waiting to be filled with science and data. There's no greater, no more mysterious, no more misunderstood organ in our bodies than our brains. If one weighed the pages of mythology around the brain against that of all scientific papers ever written about it, the scale would likely tip towards myth.

---

43 *http://www.dan.sperber.fr/wp-content/uploads/2009/10/MercierSperberWhydohumansreason.pdf*

The fields of psychology and neuroscience are filled with misinformation, disagreement, untested hypotheses, and the occasional consensus-based, verifiable, and repeatably tested theory. And so it's a struggle for me: on one hand, I'm preaching about information diets, but—in trying to synthesize my own research in the field—I run the risk of accidentally feeding you junk information myself. On the other hand, so much of both fields is applicable to an information diet that it's impossible not to draw on them.

Banting had an advantage on me. When he wrote his *Letter on Corpulence*, the Calorie was a unit used to measure the energy consumption of steam engines. Science had not scratched the surface of what he'd touched on yet. I'm writing in the midst of the dawn of science in this field; we know some, but not a lot. It's as scientifically accurate to say, "This MSNBC is going straight to my amygdala," as it is to say, "This ice cream is going straight to my thighs." Only, now we actually have more information and more accurate research about how ice cream actually affects your thighs.

Let's start with some acknowledgement that our brains are not exactly like the digestive and endocrine systems. Direct comparisons tend to be ridiculous: the rules for how our minds store and process information are different from how our bodies store and process food. Food consumption has immediate effects: drink an extraordinary amount of water, and you may get a fatal case of water intoxication. The same is not true for information; few people have died directly from reading too much *PerezHilton.com* in a given day.

Cognitive processing does, however, cause physiological changes just like our food does—only not in the same way. Up until a few years ago, it was thought that the human brain became fixed at some point during early childhood. Now science has shown that this isn't the case; our brains constantly adapt and change their physiological structure. Every time we learn something (according to neuroscientists), it results in a physiological change in the brain.

This phenomenon is called *neuroplasticity*, and a quote from Dr. Donald Hebb sums it up: "neurons that fire together, wire together." More explicitly, Hebb says:

"Let us assume that the persistence or repetition of a reverberatory activity (or "trace") tends to induce lasting cellular changes that add to its stability…. When an axon of cell A is near enough to excite a cell B and repeatedly or persistently takes part in firing it, some growth process or metabolic change takes place in one or both cells such that A's efficiency, as one of the cells firing B, is increased."[44]

The human brain is constantly adapting to experiences and the choices the mind makes. In London, taxi drivers must pass a comprehensive exam known as "The Knowledge," which requires them to instantly create routes for passengers without the use of a GPS or a map. It's considered the world's most comprehensive taxi driver's test, and it takes up to four years to prepare for and pass it.

According to scientists at the University College London, this is why London cab drivers have a differently shaped hippocampus than "regular" people.[45] The hippocampus is important to the brain's ability to move short-term memories to long-term memory and to help with spatial navigation, the skills the cab drivers in London need the most. As a cab driver exercises this part of the brain more, the brain adjusts and lends more neurons to the region. When that happens, old circuits are replaced by new ones.

That's one example of how doing things changes the physical composition of the brain. What about just reading something? Could something with a lower cognitive load—like watching your favorite television program—alter your brain's structure?

The answer is likely *yes*. Every time you learn something new, it results in a physiological change in your brain. In 2005, in the paper "Invariant visual representation by single neurons in the human brain," Quiroga et al. found that a single cell in the human brain fired off only when a picture of Jennifer Aniston alone was shown to a test subject. Another, distinct neuron showed up when the subject was shown a picture of Halle Berry, another for a picture of the Sydney Opera House, and another for the Baha'i temple. [46]

They even found a correlation between the neuron that fired off when a picture was viewed of the common landmark or celebrity, and the string of letters representing the corresponding name. In other words: a picture of the Sydney Opera House fired off the same neuron that seeing the string "Sydney Opera" did.

---

44 *http://www.nature.com/?file=/neuro/journal/v3/n11s/full/nn1100_1166.html*

45 *http://news.bbc.co.uk/2/hi/677048.stm*

46 *http://www.vis.caltech.edu/~rodri/papers/nature03687.pdf*

But that's about memory. What about beliefs?

Dr. Ryota Kanai and some colleagues at the University College London Institute of Cognitive Neuroscience took self-described liberals and conservatives and studied their brains via fMRI (functional magnetic resonance imaging). They found something remarkable: the liberal brains had structural differences in the anterior cingulate cortex—the region of the brain responsible for empathy and conflict monitoring.[47] Conservative brain structures, by contrast, had enlarged right amygdali—the part of the brain responsible for picking up threatening facial expressions and responding to threatening situations aggressively.

The science is admittedly sketchy. Kanai's test group was rather limited— just a small group of students. It also doesn't explicitly prove that environmental factors had anything to do with the increased sizes of the respective brain regions; this could be genetic.

That said, I contacted Kanai and asked him if these differences in brain region sizes could be the result of media consumption and other environmental factors. Here's what he had to say:

> *"From our study, it's hard to resolve the chicken or the egg causality of brain structure and political orientation. I think this needs to be further explored with additional empirical work. As you suggested, exposure to politically tinged information could have influenced people's political opinions, and it would be very interesting to see if such changes are reflected in brain structure. This is an empirical question we have to answer by more experiments."*

I also contacted another respected neuroscientist in the field, Dr. Marco Iacoboni at UCLA, to see what he had to say:

> *"I think it's plausible, although unprovable at this stage. I mean, any decision we make is based on neurophysiological activity, it doesn't come from the gods. If people, on average, become more or less liberal, in some way something must have happened in their brain. The tricky issue is the chain of causes and effects."*

---

47 *http://www.sciencedirect.com/science?_ob=MImg&_imagekey=B6VRT-52JV2HC-2-1&_ cdi=6243&_user=10&_pii=S0960982211002892&_origin=gateway&_coverDate=04/26/2011&_sk =%23TOC%236243%232011%23999789991%233131734%23FLA%23display%23Volume_21,_ Issue_8,_Pages_625-718,_R267-R290_(26_April_2011)%23tagged%23Volume%23first%3D21%2 3Issue%23first%3D8%23date%23(26_April_2011)%23&view=c&_gw=y&wchp=dGLzVtb-zSkWb& md5=b8fd284296848eb5f1af268e7a4da819&ie=/sdarticle.pdf*

Whether or not media consumption could physically alter your brain to be more partisan is unknown. But what's known is that whenever you learn something new, the result is a physiological change in the body—just like whenever you eat. Another similarity is that we're not in direct control over what changes get made to our brains, and where.

Search Frenzy

Back in 1954, psychologist James Olds found that if he allowed a rat to pull a lever and administer a shock to its own lateral hypothalamus, a shock that produced intense pleasure, the rat would keep pressing the lever, over and over again, until it died. He found that "the control exercised over the animal's behavior by means of this reward is extreme, possibly exceeding that exercised by any other reward previously used in animal experimentation."[48] This launched the study of brain stimulation reinforcement, which has been shown to exist in all species tested, including humans. At the heart of brain stimulus reinforcement is a neurotransmitter called *dopamine*.

Dopamine makes us seek, which causes us to receive more dopamine, which causes us to seek more.[49] That jolt you feel when you get a new email in your inbox, or hear the sound of your cell phone's ding? That's dopamine, and it puts you in a frenzy. This used to be helpful: our dopamine systems helped us, as a species, to find resources, acquire knowledge, and innovate. But in an age of abundance, there are new consequences.

Dopamine receptors often put us in a loop. With all the inputs available to us today—all the various places where notifications come about: our email boxes, our text messages, our various social network feeds, and blogs to read—our brains throw us into a runaway loop in which we're not able to focus on a given task at hand. Rather, we keep pursuing new dopamine reinforcement from the deluge of notifications headed our way.

We got this way because of evolution. We're wired to seek. For thousands of years, those that sought information got to live longer, got to have sex, and pass on their genes. We're information-consumption machines that evolved in a world where information about survival was scarce. But now it's abundant. With cheap information all around us, if we don't consume it responsibly, it could have serious health consequences.

---

48 *http://www.wadsworth.com/psychology_d/templates/student_resources/0155060678_rathus/ps/ps02.html*

49 *http://www.sott.net/articles/show/222397-Dopamine-Makes-You-Addicted-To-Seeking-Information*

CHAPTER 5

Welcome to Information Obesity

"I remember when, I remember
I remember when I lost my mind
There was something so pleasant about that place
Even your emotions have an echo in so much space
And when you're out there without care
Yeah, I was out of touch
But it wasn't because I didn't know enough
I just knew too much
Does that make me crazy?
Possibly"

—Gnarls Barkley, "Crazy"

In 2004, after two years of my Howard Dean–based information diet, I headed home to Albany, Georgia, for Thanksgiving dinner. It's like a national holiday in my giant southern family—come to my family's house on Thanksgiving and you'll begin to wonder whether or not the whole thing about us all being related in the South is true. It seems like the whole state is at our house.

One of my favorite relatives is my dad's brother—Uncle Warren. He's got a deep drawl that makes you want to sip lemonade and sit on the front porch all day, and is so charming that he could convince Alex Trebek to give up on the trivia questions and just go fishing. He also loves Fox News, and thinks that if we all listened to a little more of it, we all might be a bit better off.

In that fall of 2004, Uncle Warren and I had very different information diets. It had become a sort of family tradition in the years prior for my uncle and I to clash, but this year I made it a point before my voyage home, that Uncle Warren—the symbolic head of the ultra-right-wing sect of the Johnson family—would be proven unequivocally wrong on all issues.

In my mind, I was right, and he was wrong. It just had to be the case. I mean—every news report I'd ever read on DailyKos, every data point I'd seen, and all the polls I'd seen on Zogby agreed with *me*. If I just explained the "facts" to him, he'd have to admit that, well, he was wrong and, more importantly, I was right.

What I didn't account for was that Uncle Warren watched the news he believed in, too. He showed up with his own set of "facts." I'd also neglected to read the aforementioned studies about how facts tend to be particularly poor at persuasion, especially the ones that you pick up on left-wing blogs or the ones that you pick up in the left-of-center bars in Washington, D.C.

The conversation, viewed from the outside, couldn't have been considered coherent. It was just an explosion of nonsense—two grown men shouting at each other about taxes, gun control, and healthcare, with some perfectly good turkey between us. That Thanksgiving, Uncle Warren left early, and things haven't been the same between us since.

It's a shame, really. My uncle and I in a big fight, essentially invading a family tradition because we had developed our own biases, and had both become too attached to our separate—and separating—ideas of reality. Information obesity can inflict some pretty serious damage on families.

Our media companies aren't neuroscientists, nor are they conspiratorial. There's no elaborate plot aimed at driving Americans apart to play against each other in games of reds vs. blues. A more pragmatic view is that our economy is organic. Through the tests of trial and error, our media companies have figured out what we want, and are giving it to us. It turns out, the more they give it to us, the more we want. It's a self-reinforcing feedback loop.

When we tell ourselves, and listen to, only what we want to hear, we can end up so far from reality that we start making poor decisions. The result is a public that's being torn apart, only comfortable hearing the reality that's unique to their particular tribe.

Even our leaders in Congress feel the effects. In September of 2011, Speaker of the House John Boehner said that sometimes talking to the President was like "two different people from two different planets who barely understand each other."[50]

It's a new kind of ignorance epidemic: *information obesity.*

---

50 Fox News: O'Reilly Factor, July 22, 2011.

According to a 2009 Public Policy Polling survey, 39% of people in the United States believe that the United States government should stay out of Medicare (a government-run program). Ten percent of people are not certain or do not believe that Hawaii is a state, and 7% believe that Barack Obama is from France.[51]

We know that we're losing touch, too. Or at least we have a feeling that we're losing our grasp of the facts. In a December 2010 poll done by *WorldPublicOpinion.org* (run by the Program on International Policy Attitudes at the University of Maryland), 91% of voters said they had encountered misleading information. Fifty-six percent said they'd encountered it frequently. According to Gallup, only 27% of us have a "great deal of confidence" in television news broadcasts—down from 46% in 1993.

These problems don't stem from a lack of information. They stem from a new kind of ignorance: one that results in the selection and consumption of information that is demonstrably wrong. We don't trust "the news" but we do trust "our news," in other words, the news we want to believe in. And that's a far more potent weapon than our classic view of ignorance.

In 1969, the tobacco company Brown & Williamson came to the conclusion that the only way it was going to save itself from intense regulation and market loss due to health concerns over cigarette use was through the *production of information*, rather than the concealment of it. Here's what it said:

> *"Doubt is our product since it is the best means of competing with the 'body of fact' that exists in the mind of the general public."*[52]

With that, Big Tobacco started producing as much doubt as it did cigarettes. It spent millions of dollars on funding research, and academic papers were released saying that Alzheimer's Disease is prevented by smoking cigarettes (makes sense: you'll be dead before you get it), and that smoking may boost immune systems.

Big Tobacco created independent organizations, too, such as the Center for Indoor Air Research to "broaden research in the field of indoor air quality generally and expand interest beyond the misplaced emphasis solely on environmental tobacco smoke." It set up Associates for Research In the Science of Enjoyment (ARISE) to provide research bolstering the idea that smoking's effects on relaxation could boost immune systems.

---

51 *http://www.scribd.com/doc/36367943/PPP-Release-National-819513*
52 *http://tobaccodocuments.org/landman/332506.html*

Big Tobacco figured something out: it didn't have to worry about people who smoked and didn't watch the news. They'd keep smoking. It needed to create doubt in the minds of smokers who *do* watch the news to keep them as customers, and it would need to create doubt in the minds of the non-smoking public in order to keep government away from the regulation of secondhand smoke.

In hindsight, we now know that cigarette smoke causes cancer, heart disease, and a variety of other terminal issues. Your health insurance company asks if you're a smoker for a reason: because if you are, you're going to die sooner, and probably cost a bunch of money before you go out. We're now so certain of it that some are suggesting that early smoking deaths could help save Social Security.[53]

The argument about tobacco is mostly settled, but we still face the production of doubt. In 2007, the American Association of Petroleum Geologists—the last major scientific body to reject climate change's existence and cause—changed its mind. Climate scientists reached consensus: global warming is "unequivocal" and mankind is the primary cause.[54] Since then, no recognized scientific body has dissented from the theory[55] or rejected the idea of climate change.

In the five years since consensus was reached by the scientific community, the number of people doubting climate change's occurrence has increased. When the battle for scientific minds ended, the doubt production machines shifted into overdrive.

In 1998, a public relations representative for the American Petroleum Institute named Joe Walker had foresight. He wrote an eight-page memo suggesting that the institute spend $5 million over two years to "maximize the impact of scientific views" consistent with theirs and noted that "public opinion is open to change on climate change."[56] Fast forward to 2007, the *Guardian* reported that the American Enterprise Institute, a think tank funded by companies like ExxonMobil and Phillip Morris, started offering $10,000 "grants" plus travel expenses to scientists who would publish articles emphasizing the shortcomings of theories of climate change.[57]

---

53 *http://www.stltoday.com/news/local/columns/bill-mcclellan/article_080530ab-b4b2-5c26-8e44-9eb2efcf5ec5.html*

54 *http://www.nytimes.com/2007/02/03/science/earth/03climate.html?ex=1328158800&en=61f423 12221df544&ei=5090&partner=rssuserland&emc=rss*

55 *http://books.google.com/books?id=PXJlqCkb7YlC&printsec=frontcover&dq=isbn:9780262541930 &hl=en&ei=uYyGTaz8PKKx0QHQguXhCA&sa=X&oi=book_result&ct=result&resnum=1&ved=0CC kQ6AEwAA#v=onepage&q&f=false*

56 *http://www.nytimes.com/1998/04/26/us/industrial-group-plans-to-battle-climate-treaty.html*

57 *http://www.guardian.co.uk/environment/2007/feb/02/frontpagenews.climatechange*

Slowly but surely, these scientists have produced pseudo and counter facts for money, and muddied the waters of public debate.

In 2009, thousands of emails and documents from the University of East Anglia's Climate Research Unit illegally found their way onto the Internet. The emails lit a fire across the Web called *climategate*, empowering climate skeptics like James Delingpole, author of "365 Ways to Drive a Liberal Crazy" to declare it the "final nail in the coffin for Anthropenic Global Warming."[58]

Former vice presidential candidate Sarah Palin said that all climate science was "junk science and doomsday scare tactics pushed by an environmental priesthood."[59] Soon after, radio host Rush Limbaugh and a variety of climate skeptic "pundits" piled on with sound bites. The leaked emails gave folks like the American Enterprise Institute along with allied pundits what they needed in order to create a closed epistemic loop that pulled a significant portion of people away from scientific truth.

Of course, investigation after investigation has shown that the emails were taken largely out of context, and the scientists doing the research were exonerated. That's according to an independent internal review, as well as others by Penn State, the British House of Commons' Science and Technology Committee, the UK's Royal Society, and the United States Department of Commerce. But the damage was done, giving those who wanted to deny climate change's existence something to point to that confirmed their beliefs.

You have to ask yourself what's more likely: that nearly every scientist in the broadest, most skeptical fields of science (everyone from pediatricians to geologists) has been co-opted by Al Gore's secret agenda to make us carpool, or that a smaller number of companies with billions of dollars invested in the status quo are manufacturing doubt so that we don't change. To me, the latter seems a lot more feasible. I've had the pleasure of meeting Al Gore, and he's just not that charming.

Confident Ignorance

In 2011, comedian Jon Stewart stated that Fox News viewers were "the most consistently misinformed" viewers of the media. It set off a bit of a controversy, and Politifact, a reputable fact-checking organization run by the *St. Petersberg Times* in Florida, jumped on the story.

---

58 *http://blogs.telegraph.co.uk/news/jamesdelingpole/100017393/climategate-the-final-nail-in-the-coffin-of-anthropogenic-global-warming/*

59 *https://www.facebook.com/note.php?note_id=188540473434*

They pointed to public polling from Pew and the University of Maryland—reputable pollsters—that found that viewers of shows like the O'Reilly Factor were actually just as knowledgeable about politics (through correct answers to questions like "Who is the Speaker of the House?" and "Who is the president of Russia?"), scored as more highly informed than average media viewers, and were in roughly the same league as viewers of the Daily Show, PBS's News Hour with Jim Lehrer, and National Public Radio.

Stewart responded by apologizing to Politifact for being misinformed, but then fired back with a laundry list of news stories that Politifact itself stated that Fox had gotten wrong: headlines like "Texas Board of Education May Eliminate Christmas and the Constitution from Textbooks" and "Cash for Clunkers Will Give Government Complete Access to Your Computer."

The truth is, they're both right, and pinpoint our new kind of ignorance: one that comes from the consumption of information, not the lack of it. The new ignorance has three flavors—all of which lead us to information obesity: agnotology, epistemic closure, and filter failure.

Agnotology

Robert Proctor is an historian at Stanford University, and the first historian to testify against the tobacco industry. Through his study, he coined the term *agnotology* to describe what Big Tobacco pushed on society in the later half of the twentieth century, and what the coal, petroleum, steel, and other industries through the American Enterprise Institute and the national Chamber of Commerce are doing to us now. He defines agnotology as the study of *culturally induced doubt*, particularly through the production of seemingly factual data. It's a modern form of manufactured ignorance.

Agnotological ignorance does not affect those who don't tune in. It affects those who do. At the University of New Hampshire, Professor Lawrence Hamilton polled 2,051 people across different regions in the United States. He asked them how informed they were about climate change, where they stood on the issue, and what their political party was.

The results shouldn't be surprising if you've read this far: those who claimed to know the most about climate change (as a result of consuming news or scientific data) had the most divergent opinions of its cause. Those who claimed to know very little about climate change were closer together in their opinions.

In 2008, Pew found a similar result around the climate change debate: 19% of Republicans with college degrees believed that global warming was happening because of human activity, versus 31% for Republicans without college degrees. Eighty-five percent of Democrats with college degrees believed that global warming was happening because of human activity versus 52% of those without degrees. The more informed someone is, the more hardened their beliefs become; whether they're correct is an entirely different matter.

Epistemic Closure

The CATO institute is a right-leaning, pro-business, libertarian think tank based in Washington, D.C. The offices of CATO are not lined with people holding signs that say "keep your government hands of my Medicare"—but rather with smart people who tend to believe that market forces can settle things more effectively and efficiently than government regulations. While they'll find more comfort at the cocktail parties on the right, these politicos tend to hang out by themselves, unable to find an intellectually honest home (or bar) in either party. Most Republicans identify with libertarians. But not all libertarians identify with Republicans.

Julian Sanchez is one of those folks from CATO who is probably too smart and too honest to get invited to too many cocktail parties on the right or on the left. In early 2010, Sanchez described a problem he saw with the right:

> "One of the more striking features of the contemporary conservative movement is the extent to which it has been moving toward epistemic closure. Reality is defined by a multimedia array of interconnected and cross promoting conservative blogs, radio programs, magazines, and of course, Fox News. Whatever conflicts with that reality can be dismissed out of hand because it comes from the liberal media, and is therefore ipso facto not to be trusted. (How do you know they're liberal? Well, they disagree with the conservative media!)"[60]

Climate change is a perfect example of *epistemic closure*: science is liberal; climate change is from science; thus climate change is a liberal conspiracy. Every news outlet that reports on it must also be corrupted by liberal influence, and thus can be dismissed. But the left succumbs to epistemic closure too.

Look at the left's unyielding relationship to organized labor: no institution with that much money is unquestionably good, yet you'll find many a left-wing operative in Washington looking at you sternly if you question a union's motives. Talk with a liberal about former District of Columbia schools chancellor Michelle Rhee's idea that teachers ought to be kept on

---

60 *http://www.juliansanchez.com/2010/03/26/frum-cocktail-parties-and-the-threat-of-doubt/*

the payrolls based on their performance, rather than their seniority, and you'll find yourself in a screaming match pretty quickly.

With its general distrust of pharmaceutical companies, the left is still listening to the likes of Jim Carey and Jenny McCarthy on the now-settled question of whether measles, mumps, and rubella (MMR) vaccination causes autism. (It doesn't.) The left still bristles at the question of nuclear energy, though for every person that dies from nuclear energy, 4,000 people die from coal production.[61]

Epistemic closure is a tool that empowers agnotological ignorance. As certain information is produced, all other sources of information are dismissed as unreliable or worse, conspiratorial.

Filter Failure

You don't need the liberal or conservative media to make you ignorant. It can come from the production and consumption of information from your friends, and the personalization of that information. The friends we choose and the places we go all give us a new kind of bubble within which to consume information. My experience of delusion on the Dean campaign wasn't just about my media consumption, but also the association with people who thought, consumed, and believed exactly as I did.

We all live in our own social bubbles, which we create and empower through our social relationships—and interestingly, new research says that these relationships have profound impacts on us. The friends we select, and the communities in which we work, play, and love serve as filters for us. It's too high of a cognitive and ego burden to surround ourselves with people that we disagree with.

If you're a Facebook user, try counting up the number of friends you have who share your political beliefs. Unless you're working hard to do otherwise, it's likely that you've surrounded yourself with people who skew towards your beliefs. Now look beyond political beliefs—how many of your friends share the same economic class as you?

With social media, it becomes more difficult to escape these biases. Eli Pariser, the former executive director of *MoveOn.org*, observed that though he worked hard to maintain strong relationships with conservative friends, the links that they were posting to Facebook suddenly disappeared from his news feed. Why? Facebook had determined that he wasn't clicking

---

61 *http://nextbigfuture.com/2011/03/deaths-per-twh-by-energy-source.html*

on them, and thus Facebook decided to remove them to make more room for the stuff that he was clicking on.

Before Pariser's Facebook feed got personalized for him, though—and before his web searches, online newspapers, and blogs were personalized by other companies—Pariser made some choices of his own: he chose his friends, he chose what to click on, and he chose how long to spend consuming that information. All of that information went in to the algorithms that predict what will interest him in the future.

Those algorithms are everywhere: our web searches, our online purchases, our advertisements. This network of predictions is what Pariser calls the *Filter Bubble* in his book by the same name (Penguin Press)—the network of personalization technology that figures out what you want and keeps feeding you that at the expense of what you don't want. It's different than the A/B testing based on popular opinion; this systemic personalization is supposed to bring us what is relevant to us.

Pariser's filter bubble existed long before the invention of personalized technologies. We started doing it ourselves when we started forming societies and developing our own personal networks. We tend to associate with people who believe the same things we do, unless we have to associate with them by force of turkey, like me and Uncle Warren.

What is new is automatic personalization as a way of coping with surplus information, and the fact that those choices we're making are having more immediate, more transparent consequences.

Personalization is just a mirror that reflects our behavior back to us, and while some might argue that the best way to make our reflections look better is to change the shape of the mirror, the fairest way to do it is to change what it's reflecting. We build filters around us with every friend we make, and every time we click. Without careful consideration, we risk throwing ourselves into more agnotological bubbles, and drifting farther away from reality.

CHAPTER 6

The Symptoms of Information Obesity

"There's nothing on it worthwhile, and we're not going to watch it in this household, and I don't want it in your intellectual diet."

—Kent Farnsworth
summarizing his father Philo Farnsworth's
view on the device he invented: the television[62]

My wife Roz is actually three people: there's Normal Roz, there's Email Roz, and there's Zombie Roz. Let me explain.

Normal Roz is a sharp-as-a-tack, sweet-as-a-pixie-stick, pretty-as-they-make-'em woman. She loves the outdoors, loves to garden, and loves to get her hands dirty. She combines a French love for life with the German love of hard work and efficiency. She's been known to say to me, whilst I'm in the midst of enjoying the miracles of central air conditioning: "Clay, yard work is just like that video game you're playing, except with a productive outcome."

But should she run across a computer screen on her way outside to try to plant corn in our 16-square-foot back yard, it's over. Especially if an email window is open. She will sit down in front of her computer, and (according to her) time no longer exists. Hours later, she'll look up at me, eyes blood-shot, and wonder why I'm asking her to come to bed. Time stands still for her, the day passes, and she has no idea where it went. Email Roz has no sense of time. I won't lie—sometimes when she wants me to go do yard work I have left a laptop open between her and the door. Works every time.

But the scary Roz is Zombie Roz. Normal Roz can be on her way anywhere, and if there's a television playing anything from Fox News to HGTV, Normal Roz turns into Zombie Roz: transfixed, and mouth agape at the

---

62 http://www.byhigh.org/History/Farnsworth/PhiloT1924.html

television. It's as though a freeze-ray shoots out of our TV, and once it enters her field of vision, she's powerless to resist it. I once watched her stare blankly for 15 minutes at a Spanish language cable network. She doesn't speak Spanish.

I've caricaturized my wife to make a point: information consumption makes you sedentary, and sometimes, it ruins your sense of time. Being sedentary is bad for your health.

The Connection Between Obesities

It turns out that sitting for long periods of time isn't particularly good for you. Whether you're sitting behind a computer, sitting in front of a television, sitting in your car listening to the radio on your way to work, or sitting and reading this book, we are usually sedentary when we're consuming information.

In 2004, one physician coined the term *Sedentary Death Syndrome* to classify all the diseases that come from the sedentary state. The effects: heart disease, diabetes, cancer, and yes, obesity. Some researchers are calling it the second largest threat to public health in America. What are we doing when we're sedentary? Few of us are meditating. We're usually consuming information.

New research points to sitting, especially amongst men, as a leading cause of death. Even if you exercise regularly, it turns out that sitting for long periods of time can be deadly. Dr. James Levine of the Mayo Clinic:

> *"Adults who spent more than four hours a day sitting in front of the television had an 80 percent increased risk of death from cardiovascular disease compared with adults who spent less than two hours a day in front of the TV. This risk was independent of other risk factors such as smoking or diet.*
>
> *And it's not just TV watching. Any extended sitting—whether that's at a desk or behind the wheel—increases your risk. What's more, a few hours a week at the gym doesn't seem to significantly offset the risk."* [63]

Most of us aren't consuming information while jogging on a treadmill. If you have a desk job, it's likely that your desk is one that comes with a chair, not a pad on the ground for comfortable standing. But as we sit there in

---

63 *http://www.mayoclinic.com/health/sitting/AN02082*

front of our computers, we are slowly killing ourselves just waiting for the next hit of dopamine to come into our inbox.

As of 2008, according to the UCSD, we were consuming 11.8 hours of information per day per person while we're not at work.[64] While some of that may be listening to music or the radio while we're running on the treadmill, most of those hours are spent sitting down.

Do us both a favor: stand up and stretch, take a break and walk around for a bit. I'd like for you to finish this book.

The nice thing about physical obesity is that you can pretty easily tell if you're obese. Body dysmorphia aside, one need only to fly on a major airline to check oneself—if you need a seatbelt extender, you're likely obese. If you cannot see, much less touch, your toes because your belly is in the way, you're likely obese. And if you don't want to try those tests, any trip to the scale will tell you whether or not you're suffering from obesity.

The dangerous thing about information obesity is that it's a bit more nefarious. It's difficult to tell if you are suffering from information obesity or have poor information consumption habits. It's impossible to know if you're ignorant and as we've learned, even if confronted with our own ignorance, it's likely only to make us run out and consume more misinformation in order to avoid being wrong.

Socrates' view on this was simple: just accept your own ignorance as the only thing to be certain about. This view is important to keep in mind, and a healthy foundation for an information diet.

Information obesity isn't new. Just as it was possible to be obese 500 years ago, it was possible to experience this new kind of ignorance 500 years ago, too. It was just more expensive, and you had to work much harder for it. But now we're living in a world of abundance, and as it turns out, information obesity has some pretty serious consequences for our productivity, our health, and our society.

Apnea

Linda Stone is a quiet, deliberate woman who is obsessed with our autonomic nervous system: the stuff that we do relatively unconsciously, like make our hearts beat, make our skin perspire, make our mouths salivate, digest food, and, to an extent, breathe. She's constantly trying to figure out how technology affects this part of our daily function.

---

64 *http://hmi.ucsd.edu/howmuchinfo.php*

At the recommendation of her doctor, Stone started doing daily breathing exercises every morning to increase her respiratory health and reduce stress. Every morning, she gets up, goes for a short walk, and does 20 minutes of Buteyko breathing exercises.

When I met with her for the writing of this book, our meeting involved a few gadgets—the emWave2 and the StressEraser, small little contraptions that, when hooked up to your earlobe or the tip of your finger, show you how well you're breathing, and what your heart rate looks like. They're pretty simple devices that use a variety of blinking lights and sounds to calm you down and help you achieve an optimal rate of breath.[65]

You can even attach them to your computer to keep a diary of your breath. Stone often sits with a clip dangling from her ear and into her computer so that she can receive constant feedback there on her screen about her breath and heart rate, and continually try to stay in a relaxed, primed state.

After a few days of these breathing exercises, she noticed something interesting: just a few minutes after doing her breathing exercises, she'd head to work, check her email, and find herself holding her breath. Noting that there may be something wrong with that, she grabbed her gadgets and got to work finding out if she was the only one holding her breath in front of a monitor.

After about seven months, and about 200 interviews, Stone found that 80% of the people that she talked to and observed were holding their breath—especially when email came into their inbox.

So I decided to buy one of these devices and test myself during the writing of this book. I scheduled email checks only twice a day for one hour, and found that during those hours, sure enough, my breathing was more shallow and more irregular than during the hours in which I was writing.

Linda describes the problem with a term she coined: *email apnea*. But the irregularities go beyond email: I found that when I was dealing with all different sorts of incoming information online, my breath and heart rate became irregular. Any time I was dealing with something with a number by it or a queue, my breathing changed.

I noticed something else interesting when I dusted off my marathon training heart rate monitor and began to wear it to the office during the day: when I received a text message, my heart rate went up slightly and wouldn't go down until I read the text message or after about five minutes—the amount

---

65 *http://www.huffingtonpost.com/linda-stone/just-breathe-building-the_b_85651.html*

PART I: INTRODUCTION

of time I suppose it took me to refocus on my work and forget about the message.

I'm uncomfortable with this method of consumption. I don't like a device giving even my most honest and caring of friends the ability to increase my heart rate with the push of a button. It's too Pavlovian for me. And holding your breath has some serious consequences; not only does it regulate the amount of oxygen and carbon dioxide in your blood, but it also helps regulate your fight/flight response. A lack of oxygen comes with a variety of awful health consequences like diabetes and obesity.

Poor Sense of Time

When Email Roz looks up at me after a six hour inbox tour, she seems disoriented. How could it possibly be so late? She was just on her way outside to do some gardening. "What do you mean it's dark outside?" she'll ask. Her eyes are a little bloodshot—it's like waking a person up.

I get it. The same thing happens to me. When I sit down in front of a computer, it is as though the world around it disappears. Metaphorical blinders go on, and it's as though I'm almost inside the computer itself. I've been captured by my 27" iMac. When I step out from of a long run in front of a computer, it's almost as though I have to reorient myself in the same way that I reorient myself in the morning when I wake up.

Every time you get a new email, text message, or other kind of notification, you also get a little hit of our old friend dopamine. It turns out that dopamine not only puts us into a seeking frenzy, but it also distorts our sense of time. We can spend an hour inside of our email inboxes when it feels like just a few minutes.

Email Roz and her husband Writer Clay have done some pretty terrible things to each other—they've left each other at train stations, been late to dinner dates, and let entire evenings pass them by while they've sat together. Just a quick check of the email when we get home can often end up in evenings entirely lost to LCD screens.

Attention Fatigue

About two years ago, I started to wonder: what the heck happened to my short-term memory? And where did my attention span go? I've written a little, pithy 140-character tweet, sent it into the universe, and in no more than five minutes received a reply. The only problem is, I've already forgotten what I wrote in the first place. I've had to go back, and look at what I said just five minutes ago to understand what the person replying to me is referencing.

Some days my brain just feels like it's in a state of frenzy, and I *need* to keep checking all the different things I need to check. There's just no time to read the academic papers or even to respond to that email that will take 20 minutes to respond to because there are so many *new* emails to read.

The new world of abundant information, as many have noted, is one filled with distraction. On any given day, many of us see thousands of advertisements cleverly designed to capture our attention. We come across scores of links on the Web, custom tailored just for us. Twitter streams across many a desktop, and Facebook's little red notification number beckons us with the details of our friends at every moment. Our emails, phone calls, and text messages can interrupt us at any second.

All of this wreaks havoc on our ability to sustain attention. Cory Doctorow points out that whenever we sit at our computers, we're tuning in to a new "ecosystem of interruption technologies."

Nicholas Carr points out in his book, *The Shallows* (W.W. Norton), that this chorus of siren songs of distraction is wreaking havoc on our brains. We've spent hundreds of years now training ourselves to pay attention to something as banal and repetitive as text (compared to the things we used to pay attention to like food and predators) for long periods of time. Carr bemoans the influence that these new interruption technologies are having on our brains, essentially wiring our brains to click on the most insistent distraction.

Attention is something that requires cognitive energy, and it's something that we must build up. You don't train for a marathon by sitting on a couch and you don't help your attention span by giving in to the temptation of every distraction that comes across your eyeballs. As information becomes more and more tailored, it becomes harder and harder for us to resist pursuing it, and our attention banks carry smaller and smaller balances.

Loss of Social Breadth

Social anthropologist Robin Dunbar, alongside several other scientists, has an interesting theory: our neurocognitive resources have a limit to the total number of relationships we can manage—and that number is somewhere between 100 and 250. Informally, the number is estimated to be 150, and it's called *Dunbar's number.*

Dunbar came to this conclusion by studying human tribes, hunter-gatherer types, and it's bound to remain relatively true today in the age of the social networks: there is a finite number of people that we can possibly care about, and while that number varies from person to person, it doesn't come close to the numbers that sit by our names on social networks like Facebook and Twitter.

If Dunbar is right, that means our ability to manage news from friends in new social networks, and to use it to enhance meaningful relationships, is limited. By succumbing to our biases and falling into homogenous groups or epistemic loops, we eliminate the social inputs that bring us news we disagree with. Strong bias for some non–conscious consumers means cutting off meaningful relationships with people we care about but may disagree with.

The overconsumption of specialized knowledge—whether it be political or technical or even sports-related—can make it so that the only thing you're capable of holding a conversation about is the thing that you've been so deeply into, and thus as your consumption of information around a particular subject becomes more homogenized, if you're not deliberate and careful, your social group too becomes a reflection of that homogenization.

Distorted Sense of Reality

Cults work because they get their members to either convert the people around them or dismiss the nonbelievers as heathens. They're methodical in their epistemic closure, first building up a new lens to view a lens through, and should someone else see the world differently, that person is either branded a heretic (which comes from a Greek word meaning "choice") by the orthodoxy, or a "dead agent" in the realm of scientology.[66] Most major cults have some way of labeling the outsider.

Rapture tends to be an excellent topic area in which to see the effects of epistemic closure, confirmation bias, and poor information diets. Evangelical radio host Harold Camping famously predicted that the world would end and Judgment Day would arrive on May 21, 2011. Two hundred million Christians were to be taken to heaven before a global earthquake would destroy the planet.

Camping's organization, Family Radio, spent millions of donated dollars on more than 5,000 billboards across the country. He, along with his devoted followers, were certain that on May 21, the world would be filled with rapture and spread the word to everyone who would listen. The world did. The media—television, radio, newspapers, Facebook, and Twitter—were filled with news of the impending rapture.

66 *http://www.apologeticsindex.org/d05.html*

Leading up to the event, a Yahoo group called "Time and Judgment," a group whose purpose was "to discuss the events that the Bible declares will unfold on May 21st, 2011," was filled with a thousand messages of people professing their faith and sharing plans for the rapture. Leading up to the rapture, the group fed off itself. It was as though people were competing to see who could have the most blind faith.

> Marco M., May 5th: "I have looked at the Biblical evidence for the Rapture and Judgment Day on May 21, 2011. It is solid, convergent, inter-locking and replicable. So, I have no doubt whatsoever that May 21 is Judgment Day. And yes, I have quit my job."[67]

> Tony V., May 7th: "I am a bus driver for NJ Transit and I get 4 weeks vacation. I took all my vacation in March. But I'm still working so I can spread the Word on my job and still have an income so I can continue to support FR with finances. After May 21 money will be useless, so I want to spend all my money in getting the Word out."[68]

> Enow A., May 19th: "I just want to use this last chance to write to you to let know how fortunate I consider myself to be part of those spreading the May 21, 2011[sic]. Fellowshipping with you has been a source of enormous blessing and encouragement against the fierce residence I am encountering. By HIS Grace we shall meet in HIS presence soon. God bless you all. Greetings from Cameroon."[69]

May 21st was a tough day for Mr. Camping and company. The *New York Times*' headline read on the front page: "Despite Careful Calculations, the World Does Not End." The believers had to face up to the facts: Mr. Camping's prediction, and all of their certainty, had gone down the tubes. But did they admit defeat and pack up their bags?

> Ervinclark24, May 22nd: "How significant is it that this verse Joshua 10:13 tell us that the earth is essentially a day behind? Is this not saying that yesterday, May 21 was really May 20th?"[70]

No. For these zealots, it first meant searching for reasons why Mr. Camping must have been right. It was a time zone issue, or maybe there's another Bible verse? On May 22, Mr. Camping simply proclaimed that the rapture

---

67 *http://groups.yahoo.com/group/TimeandJudgment_May212011/message/8269*

68 *http://groups.yahoo.com/group/TimeandJudgment_May212011/message/8289*

69 *http://groups.yahoo.com/group/TimeandJudgment_May212011/message/8582*

70 *http://groups.yahoo.com/group/TimeandJudgment_May212011/message/8789*

on May 21 was a spiritual judgment, and that the actual end of the world would happen on October 21.[71] Supporters were relieved.

> G Agate, May 26th: "It was a Spiritual Judgment that took place on May 21. This may take a bit of time for it to sink into our human minds, Spiritual truth does not usually come quick or easy from the bible. But the main point and purpose of the day did come to pass, and most of us all were allowed to think of other things relating to it from the bible, in a literal way, so that we would get the message out to the whole world."[72]

> Tony V., May 29th: "I love [Family Radio] and [Harold Camping] I learned a lot from his teachings, and I am praying for him. And I still believe that Oct.21 will be the last day, only because I personally checked out the time line of history... Now I can understand what the Bible says(No man knows the DAY OR HOUR OF CHRIST RETURN) but we can know the year. I send my love and prayers to all the brothers and sisters on this web site and through out the world."[73]

In the group, there were a handful of messages questioning their faith that judgment happened.

> Britton95624, May 24th: "It seems as if we are hastily jumping to conclusions about all of this without having real biblical support for any of it. Just grasping at anything. Its almost as if people are saying 'well then it must have been a spiritual judgment because we can't be wrong.' That seems like pride creeping in. We are ALL confused right now, and I hope we aren't beginning to trust in our understanding over and against God. My dad said to me last night that, Solomon was the most wise person in the world. God clearly used Solomon for many wonderful things. But in the end of his life, he exhibited behavior that was not becoming of a child of God. Wisdom in a sense can be like money. It's not bad in itself, however when placed into the hands of a man in large quantities, we may not be able to handle it. It can bring you down."[74]

> Raynakapec, May 25th: "I cannot bring myself to listen to FR anymore. I am sick at heart imagining how the dear people feel who put their beloved pets down, out of their love for them, in response to HC

---

71 *http://groups.yahoo.com/group/TimeandJudgment_May212011/message/9171*

72 *http://groups.yahoo.com/group/TimeandJudgment_May212011/message/9164*

73 *http://groups.yahoo.com/group/TimeandJudgment_May212011/message/9338*

74 *http://groups.yahoo.com/group/TimeandJudgment_May212011/message/9126*

adamantly saying that after May 21 it would be hell on earth. He repeatedly said 'It's going to happen.' I seem to remember someone asking him on OF whether the caller should put her pets down, and HC as best I can remember said 'Do what you feel best,' or something to that effect. What hell they must be going through."[75]

What hell indeed. The posts kicked off a flood of replies in the group, all—though sympathetic—assuring the doubters and dissenters that the answer was to pray, and wait for the real judgment day on October 21, 2011.

If you're reading these lines, the end of the world hasn't happened.

The stories of Mr. Camping and his followers are severe cases of reality dysmorphia. These people aren't classically ignorant. Most of them have scoured the Bible, and probably read it more thoroughly than your average church attendee. What's different is that they've picked up a bias, sacrificed something for it—their time, their money, or even their dogs—and now they're vested in it.

Brand Loyalty

Any human with an active and alert mind can fall prey to epistemic closure. There are plenty of less extreme examples to point to besides evangelical doomsdays.

You can see the same fervor in the eyes of political activists. Look in the eyes of a Code Pink supporter on the left, or someone looking for Barack Obama's birth certificate on the right, and you'll see the same kind of radical devotion to what they want to believe over the facts—and you'll also likely find that most of their social network is comprised of people who feel the same way.

Brand affiliations work this way, too. Attend a major corporate developer conference like Apple's WWDC, Google's I/O, or Facebook's F8, and you'll find the latest technologies and advances from these companies paired with sermons in the form of keynotes not just telling you why their software is the future, but why the competition's values are *wrong and misleading*.

Attending Google I/O with my iPhone was a mistake. People looked at it, scowled, and scoffed. If I tried to explain that it was an older model—one that was out before any Google phone had been released—and that I was still on a two-year contract for the subsidized phone, and couldn't switch carriers, it didn't matter. Eyes rolled.

---

75 *http://groups.yahoo.com/group/TimeandJudgment_May212011/message/9063*

A few weeks later, I sat outside Apple's WWDC with the HTC Evo 4G (a Google-powered phone) to see what would happen. Again with the remarkable judgment over something so foolish as a phone—and coming from someone wearing socks with sandals!

Having attended the keynotes from both companies, I can see why the attendees of the conferences thought that way. For them, this wasn't about the use of a phone. This was about the triumph of good over evil. Through the lens of a charged up Googleist, I was but a poor infant letting Apple decide what was good for me. To the Appleist, I was dumb enough to fall for Google's corporate messaging.

It's *West Side Story*. About phones.

Symptoms and Severity

These are all symptoms I've faced or observed in my own life as a result of information consumption, but it's certainly not an exhaustive list. There's also research on Internet addiction, screen addiction, and a variety of other addictive disorders that come alongside information overconsumption.

It's likely that if you picked up this book, then you're suffering from some of these problems, and may not realize that you're suffering from others. Though they're all frightening, they, along with a slough of social problems, aren't the real case for going on an information diet. The real case is the incredible benefits. Just like a healthy physical diet and exercise can help you live a longer, happier life, an information diet can contribute to the same, as well as more meaningful, tangible relationships with the ones you love.

PART II

The Information Diet

To me, Ed had superpowers. Ed sat in the office next to me when I was working at the search engine company Ask Jeeves about a decade ago, and I was always envious of his ability to stay healthy. In my mind, Ed did some form of triathlon that involved riding a bicycle underwater while carrying a backpack filled with sharks, and did most of his work as a product manager for Jeeves' business division while doing handstand push-ups. He is one of the healthier people I've met, and while his exercise regimen was part of it, it was his attitude about food that gave him his edge.

My favorite thing about Ed was his total contempt for carbohydrates. At lunch, if he managed to get served a biscuit as a side item for something he ordered, he'd scowl at that biscuit until it went away (usually by way of me) like it was some form of dirty filth that had invaded his tray.

That biscuit wasn't there on his plate to tempt him. It was there to kill him: a little, fluffy, white, buttery enemy waiting to pounce at any moment. But did he throw away his biscuit? No. Then he wouldn't be able to keep an eye on it, lest it try and escape. As though he was persistently testing his will, Ed would keep the biscuit on his desk to sit and grow stale as a frequent affirmation that he didn't need that pile of empty carbs.

In the world of information, there are thousands of biscuits all around us waiting to be eaten. It's up to us to choose whether to chow down or to stare at them with contempt.

Imagine a world where liberals stare at Keith Olbermann's show in the airport, not eagerly awaiting a confirmation of their beliefs, but in contempt for what these shows really are: biscuits in broccoli's clothing. Or conservatives asking to shut off the O'Reilly Factor at the bar because it may ensnare them in a closed epistemic loop. Or Apple fans going on a gadget blog fast for the weeks surrounding the latest iPhone announcements in order to make a more rational decision about how to spend their next $600. Those are the decisions that people who are trying to have healthy information diets make. We should be staring at these dopamine delivery services with as much contempt as Ed does his biscuits.

The result of going on a healthy information diet is better health and a better life. The next few chapters introduce the concepts and the framework for achieving that. They're designed to build the literacy and skills required to do it, and include recommendations for the habits it takes to consume and live well in a world of abundant information.

You'll have more time to do things you enjoy, and you'll spend less time doing the things you don't. You'll also likely live longer by reducing the things that cause you stress or cause you to make poor decisions. Because you'll be consuming critically and deliberately, you'll be able to even make better decisions when it comes to things like your food diet. Finally, you may be less at risk for a variety of mental health diseases like depression and a host of anxiety and mood disorders.

We need to get something straight before we jump in to what a healthy information diet looks like, though: fasting is not dieting.

It's good to disconnect—everybody needs a good vacation. But unplugging, "Internet sabbaticals," "social media vacations," and "email bankruptcies" are all ways to avoid the real problem: our own bad habits. Ask any nutritionist, and they'll tell you that a diet isn't about not eating—it's about changing your consumption habits.

Being thin isn't the point of a good diet, either; it's about a healthy lifestyle. Our obsession with weight rather than nutrition has us confused and lined up to be taken advantage of by shucksters in the bookstore promising us that we can lose weight and look just like the airbrushed people on the book's cover if we just follow their simple, easy, attainable plan.

Let's not fall into the same trap with an information diet. Just like a normal, healthy food diet, an information diet is not about consuming less; it's about consuming right. The next few chapters of the book describe a framework for building a healthy information consumption lifestyle for yourself.

There are many possible ways to do this, and the chapters that follow are but a recommendation that comes from my own personal experience, and the experiences of others whom I've interviewed.

We're now moving from the theoretical to the practical. I would love to say that much of this is backed up by neuroscience and psychology, and some of it is, but most of it isn't—it's instead based on what I've found works for me. Though if our brains can be rewired by poor information consumption habits, then one must presume that we can rewire our brains with good information consumption habits to do the opposite.

My recommendations are just recommendations. The key is to find an information diet that works for you. Pollan's "Eat. Not too much. Mostly plants" exhoration is a helpful framework but not a strict diet. You can take that, and use it to build your own food diet. Here's my rendition:

Consume deliberately. Take in information over affirmation.

The Infovegan Way

In biology, the *trophic pyramid* is a simple construct we use to think about how energy flows through the food chain. In the food world, the people eating strictly at the bottom of the trophic pyramid are called vegans—and that's exactly what we want to emulate with our information consumption. Building on that philosophy, I coined a term in 2010—*infoveganism*—and started a blog called *Infovegan.com* to describe this lifestyle. Infovegans try to emulate the consumption habits and ethical habits of vegans in the world of information.

I'll admit: it's quite an intimidating term. A lot of people view veganism as an extreme diet, and for some, it triggers visceral reactions. Veganism is not without controversy. Even some food vegans take offense at the term, either angered at the co-opting of their name, or pointing out that the metaphor isn't perfect: lots of vegan foods are highly processed.

If you can get past the baggage that the term has, infoveganism is a valid description of what we're trying to do. Like a vegan diet, infoveganism connotes that there's more to the choice of going on an information diet than seeking a healthy lifestyle. It's also a moral decision.

At the heart of veganism is ethics. Vegans largely believe that animals, as living creatures, deserve basic moral consideration. Eating meat, they claim, has all kinds of moral implications: animal cruelty, high carbon consumption, and support of an industry without much concern for public health.

Agree with the vegans or not, you have to respect their stance. It captures perfectly what we're trying to do here with an information diet: respect the content providers that consistently provide us with good info-nutrients by sticking only to those providers, and avoiding everything else.

Like veganism, infoveganism requires conscious consumption, planning, and to a greater extent, sharpened and honed skills. To be a vegan means you've got to consistently put yourself in situations where you can maintain your diet. You cannot simply agree to go to McDonald's to grab lunch unless your diet is to consist entirely of french fries. You've got to know how to cook good-tasting vegan recipes, and know what kinds of food might be sneaking animal products in.

Being an infovegan means mastering data literacy—knowing where to get appropriate data, and knowing what to do with it, using the right kinds of tools. It means working to make sure you're not put into situations where you're forced to consume overly processed information.

It means that when you are consuming processed information, you consistently check the ingredients—if you're reading news on a new medicare proposal in Congress, it means you want to take a look at the bill itself, not just what the *Huffington Post* has to say about it.

Finally, it means a moral choice for information consumption: opting out of a system that's at least morally questionable, for a different way—a way that chooses to shun factory farmed information, politically charged affirmations—and choosing to support organizations interested in providing information consumers with source-level information and reporting that contains more truth than point-of-view.

CHAPTER 7

Data Literacy

"To invent out of knowledge means to produce inventions that are true. Every man should have a built-in automatic crap detector operating inside him. It also should have a manual drill and a crank handle in case the machine breaks down. If you're going to write, you have to find out what's bad for you. Part of that you learn fast, and then you learn what's good for you."

—Ernest Hemingway[76]

Our concept of literacy changes every time there's a major shift in information technology. Being literate used to mean knowing how to sign your name. At one point it meant the ability to read and write Latin. Today, being literate generally means being able to read and understand a newspaper in your own language.

There has always been some group of people with a closer link to the truth than the rest of society. At one point in our history, some of our ancestors had the capacity for language, and some didn't. When writing was developed, we had scribes. When the printing press was developed, the author, printer, and publisher became the new gatekeepers. After we taught everyone to read a newspaper, the journalists became the class closest to truth.

Now the problem is not a widespread inability to read and write, but the vast sea of textual, audio, and video data that we wade in every day. A new skill is necessary—one that helps filter and sort through this information.

Remember the trophic pyramid? It turns out that as energy makes its way up the food chain, its transfer gets less efficient. Consumers at each level of the pyramid convert only about 10% of the chemical energy from the step

---

76 http://www.theatlantic.com/past/docs/issues/65aug/6508manning.htm

below them on the food chain. The further up the chain you go, the less energy you get.

This is why we don't usually eat a lot of other carnivores—we tend to eat either plants or things that eat mostly plants (like cows, chickens, and pigs), but we don't tend to eat things that eat cows, chickens, or pigs (like coyotes, lions, or hawks). Agriculture can't sustain the cost it would take to transfer that kind of energy up the food chain for all of us.

In the world of information, there's a kind of trophic pyramid, too; just swap energy for truth. The further away from the source—the more secondhand or thirdhand operators there are—the less truth there is.

We learn this when we're children. We've all played a game of "operator" or "telephone," wherein one person whispers a message such as "I like chocolate" into the ear of the child next to them, who then repeats the received phrase to the next child. The message is whispered on and on between all the participants, filtered through satirists and bullies, until it comes out the other end: "Clay eats worms."

All too often, we consume information at the top of the trophic pyramid of truth, and as such, we're getting only the information that has been selected for us by a network of operators interested not in telling us the truth, but in giving us what sells. We have to move towards the base of the pyramid if we want to see what's really going on.

As we wade through ever-rising seas of abundant information, a new skill is necessary to stay near the bottom of the pyramid: the ability to process, sort, and filter vast quantities of information, or data literacy.

Don't worry, you're not going to have to learn how to be a computer programmer to have a healthy information diet—just like you don't have to be a journalist in order to read the newspaper. But the Internet is not only the best way to fill your mind with nonsense, it's also the best way to get source-level information. In order to have a healthy information diet, you must be capable of gathering information from the lowest rung in the pyramid.

Presuming you have access to a computer and the Internet, I've boiled down what I mean by *data literacy* into four major components—you need to know how to search, you need to know how to filter and process, you need to know how to produce, and you need to know how to synthesize.

You may think that you're already digitally data literate, and that you possess all of these skills. And you may be right, this chapter may be remedial for you. But I encourage even the most profoundly critical thinkers, the most brilliant statisticians, and the most talented writers to revisit what it really means to be data literate.

Search

Knowing how to effectively and efficiently use Bing, Google, or any other major search engine is now a critical form of literacy. I suspect most people reading this book will have used Google once or twice in their lives, but it's worth noting that Google in particular has a tremendous amount of information available via search besides its web index.

Knowing, for instance, that Google offers not only web search, but also the ability to search through scientific papers, patents, and laws through scholar.google.com gets you closer to the facts. And though most scholarly papers, even ones funded by taxpayer dollars, sadly sit behind paywalls, it's possible to find the title of the research paper you want to read, search for the title, and find either the document itself or a decent take on it.

Knowing Google's advanced search techniques—to search through news, blogs, discussions, and social networks, and filtering by date, time, and source—gives us a good handle on how to get the best search results.

Finally, a lot resides inside of large data repositories that aren't findable through Google. Search literacy also means the ability to find the data you're looking for outside of a search engine, and to constantly be on the lookout for these repositories. *USASpending.gov*, for instance, is an attempt by the United States government to catalog every dollar it spends on contractors. It's an incredible resource for watchdogs and budget hawks, but you won't find its data through Google search. Search literacy means being keenly aware of these kinds of sources, and constantly looking out for them.

Filter

In 2004, concerned with people's willingness to believe everything they read, a 19-year-old George Washington University student named Kyle Stoneman created a website called *gullible.info*. The site was a daily diary of fake facts like:

- As of the end of 2010, there were at least nine countries in which public flatulence was illegal. Penalties for violating the statute ranged from the equivalent of a 50 cent (US) fine to public flogging of naked buttocks to 90 days in jail.

- Approximately one-half of 1% of the annual worldwide output of the greenhouse gas carbon dioxide is due to soft drink carbonation. Despite the availability of a nearly no-cost switch to nitrogen, soda manufacturers are thus far refusing to make the change.

- Prior to the discovery of antibiotics, horse urine was commonly used to treat pink eye.

- Contrary to popular belief, cats don't actually sleep; their muscles relax and their breathing slows while the brain stays completely alert.

- On any given weekday, nearly 16% of the crowd outside the Today Show has been convicted of at least one misdemeanor. Only 3.5% have been convicted of at least one felony.

Over the course of a few months, the site became immensely popular, and sometime in 2005, an overeager Wikipedia editor took one of *gullible.info*'s pieces of trivia: "LSD guru Timothy Leary claimed to have discovered an extra primary color he referred to as 'gendale'" and added it to Mr. Leary's living historical record. Shortly thereafter, *The Guardian*—the newspaper with the second largest online readership of any English language newspaper in the world—published:

> *"He exhorted America to 'turn on, tune in and drop out' and claimed to have discovered a new primary colour—which he called gendale. Now Timothy Leary, the eccentric spokesman of the 1960s counter-culture, is to become the subject of a Hollywood movie."*

A "fact" was born, and despite Stoneman's petitions, it remained online without correction for three months. Fortunately for us, Stoneman's purpose is social experimentation and humor—looking at his Wikipedia user page, it's clear that he spends a significant amount of time clearing out and correcting *gullible.info* entries on *Wikipedia.org* for the greater good.

Gullible.info is just a small example of what someone can do on a low budget to inject ignorance into culture. The fictitious color gendale went from Stoneman's site through Wikipedia's editorial process, made it through *The Guardian*'s fact-checking process, and stayed there for three months. The only person not presenting the discovery of gendale as fact in this scenario is the source.

Search alone won't help if we're unable to find the most reliable and accurate sources of information, or we're unable to draw accurate conclusions from the data we've found. We also must be able to think critically about the information we've received, and use the best tools we can to process the information effectively. The Internet is the single biggest creator of ignorance mankind has ever created, as well as the single biggest eliminator of that ignorance. It's our ability to filter that eliminates the former and empowers the latter.

We must judge good sources through filters such as: what is the intent of the author? Is it to inform you, or is it to make a point? How does the information make you feel? Is your intent in consuming this information to confirm your beliefs or find the truth? Are you capable of viewing the information objectively?

The John S. and James L. Knight Foundation, partnered with the Aspen Institute, provides a good overview of critical thinking skills. In the Knight Commission's report[77] (available for free online, and a good read if you're interested), they describe this skill as the ability to determine "message quality, veracity, credibility, and point of view, while considering potential effects or consequences of messages."

But the skills to apply these kinds of filters alone aren't enough. Data literacy must also include the ability to do something with that raw information—to process it in some way. In an era where spreadsheets help us to make the grandest of decisions, we must have basic statistical literacy and fluency in the tools that allow us to make sense out of numerical data, not just words and ideas.

Understanding how to use a spreadsheet like Microsoft's Excel, Apple's Numbers, or Google's Spreadsheets will help you sort through and see the facts better. There is also a variety of other tools that move beyond the spreadsheet that make it easy to sort through information; these include Google's Fusion Tables, *Socrata*, and *Factual*. They take time and patience to learn, but when coupled with the enormous amount of public data that's now available online, they give us incredible new opportunities to start seeing our world more clearly through the lens of data.

Creation

Data literacy also means the ability to communicate and exchange information with others. Knowing how to publish information and the ability to take feedback are both critical skills necessary for data literacy. Tools like Blogger, Wordpress, and Typepad, and the technologies that power them, like HTML, CSS, and JavaScript—these aren't just tools for keeping a personal diary; they're tools critical to digital literacy and expression.

Content creation and digital self-expression, through the creation of text, audio, or video content, are critical components of a healthy information diet. Content creation and publication are a critical part of literacy because they help us to understand better what we say, both through the internal reflection it takes to make our findings comprehensible to others, and

---

77 *http://www.knightcomm.org/wp-content/uploads/2010/02/Informing_Communities_ Sustaining_Democracy_in_the_Digital_Age.pdf*

through the public feedback we get from putting our content in front of others.

The creation of this book—the writing and editing of it—has given me more clarity on the message within it. Many paragraphs have been tested: I've taken paragraphs that I thought may be controversial, copied them into Google+ and Facebook, and pursued dialog with those people who were willing to engage with me. It's helped me strengthen some of my arguments, see things more clearly, and more importantly, recognize when I'm being nonsensical.

Synthesis

The year 1999 was probably the most anticipated in nerd history: the year George Lucas returned to debut the latest movie in his *Star Wars* franchise— *Episode I: The Phantom Menace*. I was down in Albany, Georgia when it came out, my dad, my cousin Wallace, and I were all headed to see it.

After the movie was over, Wallace and I leapt into a discussion about the movie, praising the special effects and expressing our overall annoyance at Jar Jar Binks, now one of the least-liked movie characters in cinematic history.[78] We discussed the plot line and how we thought the other two announced movies would go.

My dad, a therapist for over 40 years by then, broke in. "I don't understand why it had to be so violent. It seemed to me like they had a forum, and all kinds of structures in place for conflict resolution. Why did people have to keep going on attacking one another when they probably could have just sat down and talked it out?"

Wallace and I rolled our eyes, and Wallace quipped: "It's called Star *Wars*, Uncle Ray. Not *Star Dialog*. That'd make for a boring movie."

Dad was right, though. In an ideal world, we'd all strive for the great synthesis of ideas, and it's a shame that more of us are concerned with winning an argument than we are getting the best out of one another. The problem is, *Star Dialog* just wouldn't make for a particularly entertaining movie (though I have to wonder if it would have been any less entertaining than *The Phantom Menace*).

The last component of data literacy is *synthesis*. Once we retrieve information, filter it, and publish it, we must be able to synthesize the ideas and concepts of others back into our ideas. Synthesis isn't entertaining,

---

[78] I have no data to back myself up on this, and the data probably doesn't exist. But trust me, Jar Jar ruined childhood memories of *Star Wars* for millions of people.

and we'd all rather argue or be entertained. View publication as a chance to get feedback and a chance to make your ideas and thoughts better—an opportunity for education as much as an opportunity to educate.

These critical thinking and data literacy concepts aren't skills you'll learn simply from reading the pages in a chapter of this book. It's a skill that takes years, lots of practice, and constant refinement to develop. But just like a regular workout, these skills are good for your health—they'll keep you living a richer, fuller life.

CHAPTER 8

Attention Fitness

My friend Rajneesh and I plodded down the pathway next to the C&O Canal on our 18th mile of a 20 mile run. For the first two hours, we talked about work or technology, politics or how our weeks had been, but after the first 10 miles our minds began to care less about the weekly minutia of our lives and more about our ability to survive. We'd been at it for about four hours, and about two hours before, we'd lost our ability to carry on cogent conversation.

I called out, "15613," and he called out, "15626," and then I replied, "15639." For the past two hours, we'd simply been adding thirteens together. We'd done this little ritual over 1200 times—about 10 times a minute—for one reason: to keep our minds focused on something else other than agony.

Raj and I had stumbled upon what scientists call a *strategic allocation of attention*. It's something that psychologist Walter Mischel discovered in 1972, when he conducted a study on deferred gratification called the "marshmallow test." [79]

Mischel brought children aged four to six years into a room free from distractions, and asked them to choose a treat: a pretzel stick, an Oreo cookie, or a marshmallow. Their treat of choice was placed on a table with a chair in front of it, and the children were told that they could have the treat right away or, if they waited for 15 minutes, they could get a second treat.

If you've ever been around kids, sweets, and willpower, you already know what happened: most of the kids fell apart. Only a third were able to double their payoff. The rest of the kids ate the marshmallow, and most didn't make it anywhere close to 15 minutes.

What Mischel found, though, was that the children who were able to pass the test wanted the marshmallow just as much as their short-term investing

---

79 *http://www.newyorker.com/reporting/2009/05/18/090518fa_fact_lehrer*

peers. What they possessed wasn't willpower, but a better skill at "strategic allocation of attention." The kids that succeeded spent the 15 minutes doing something other than obsessing about the marshmallow: they sang songs, took naps, and avoided even looking at the marshmallow. You can find the research videos on YouTube, where you'll see this effect in action if you watch. The successful children are the ones who fill their minds up with things other than the deliciousness of marshmallows.

The interesting part came a decade later, when Mischel followed up with the teenagers. The third of children that did succeed turned out to have scored higher on the SAT (Scholastic Aptitude Test) than those that couldn't wait for their sugar high. They were arguably on their way to more life success than the ones who had failed. Our ability to exercise this strategic allocation of attention is a cognitive resource that indicates academic success.

Willpower

A few years ago, I found myself completely unable to read more than a thousand words. There was no way I could read long-form journalism or even a book. The concept of reading a book, much less writing one, was completely foreign to me.

With emails to check and reply to, I could spend the entire day trapped in a sea of distraction, having accomplished nothing. My life was littered with notifications. The little email envelope icon sitting next to the clock on my computer, the Twitter notifications, and Facebook took so much time to process that I wasn't able to accomplish much else. My suffering was coming from a lack of will to focus.

Some scientists certainly seem to think it's the case that willpower is an exhaustible resource in the mind. In the book *Willpower* (Penguin, 2011), Roy Baumeister and John Tierney describe it as one of two consistent traits in people who have positive life outcomes—the other being intelligence.[80]

Their book catalogues experiments in which participants who complete a task involving their will (like resisting fresh cookies) struggle to complete completely unrelated tasks (like solving a geometric puzzle) later. Resisting a candy bar may weaken your resolve in a high-pressure sales environment like buying a car.

Willpower is part of what cognitive scientists call *executive function*. And executive function can be trained. Exercise is a healthy diet's most

---

80 There are obviously other factors that go into success and happiness. But these two are relatively constant from person to person.

important partner. I view attention, the conscious kind of focus that we all desire to be more productive, as a form of athleticism. Like running a marathon, our ability to focus depends as much on our will as it does our natural ability.

If we are training our brains to shorten our attention spans and tune in to the cacophony of distractions around us, then we must certainly be able to train it to do the opposite, and strengthen it the other way around.

Over the past few years, I've developed a framework for myself that has helped me increase my attention span. It's geared towards people who spend most of their work time behind computer screens, but the theory can be applied to all kinds of careers, and it doesn't need to be tied to work at all. It's simply a system for measuring and lengthening your attention span. What you pay attention to is a completely different matter.

Measurement

If will is a trait, like intelligence, that improves our lives, and it's an exhaustible resource, then we have to think of our attention like a currency. Our language is well-suited for this already—we don't "burn" attention, we usually "pay" it, and often times, like the federal budget, it has deficits. Our attention is the currency that marketers lust for, and it's about time we started guarding it, consciously, like we guard our bank accounts.

Nothing can be increased if it first cannot be measured. In order to track your progress, the foundation of our system relies upon good, practical measurement. We need software that can measure how we're using our computers and what we're focused on. Fortunately enough, there's great software for this called RescueTime, and it's available for the Mac or the PC. You can find it by visiting *RescueTime.com* or by visiting the resources section of *InformationDiet.com*.

RescueTime sits in the background, whenever you're using your desktop, and tracks what you pay attention to. It's a silent, impartial judge that watches every website you visit, and every window you have open on your desktop, and measures how productive you are.

During your first week using RescueTime, log in to the RescueTime website frequently to fine-tune the software. You can set up lists of websites that are healthy and necessary for you to do your job and part of your ongoing set of work, and other websites that are distractions.

Be strict with yourself: if you've found yourself constantly clicking the refresh button in your web-based email client, go ahead and mark email as antiproductive for you. Same goes with the news sites and blogs that you read. If you're an overshopper, make sure RescueTime knows that *Amazon.com* is bad for you.

Every week, RescueTime will send you an email giving you a productivity score, and comparing your productivity to that of the entire RescueTime community. What we want to do now is make this number go up.

Elimination

After you've set up RescueTime to measure your progress, you need to take a hard look at your computer and start eliminating the things that are distractions. You want to move yourself from a reactive model of computing, where you're constantly being tugged and pulled in every direction and responding to every notification that comes across your screen, into a conscious model, where you're in complete control of what you're paying attention to.

Take a look at your workspace, and silence everything that's set up to notify you of anything. Silence your phone, put it on vibrate, and put it on something soft so that you can't hear it when you're working.

Take a look in the "system tray" of your computer—either that spot near the bottom where your clock is on Windows, or that spot near the top on your Mac. If any of those little icons there (besides the clock) change color, create little cartoon bubbles, or otherwise generate notifications, get rid of them.

Close down your desktop Twitter client, and shut off your instant messages. Change your Outlook preferences to only receive new messages when you click the send and receive button.

One way to do this in modern operating systems is to create a new user account on the same computer you use, but without access to all the software that keeps you distracted. That's how I've set myself up—I have one user called "Work" and another called "Play." This gives me a container that I can put my mindless web surfing habits into, and another kept free from distraction.

Turning these interruption technologies off isn't enough, though. You'll also need to arm your web browser with tools to help eliminate distractions while you're trying to surf the Web. You can't very well be expected to accomplish a Google search for valuable information when, if you're a member of Google's social network, Google+, there's a bright red notification bar sitting there waiting to be clicked on.

Fortunately, there's a browser extension for Google Chrome and Firefox to rid you of many of the Web's distractions. On *InformationDiet.com*, I've catalogued many of them for you—but I'm certain I'm not going to be able to keep up with the ever-expanding universe of interruption technology. So here's a simple rule of thumb to live by: *if it has a number by it, eliminate it.*

Let's go ahead and get rid of those advertisements on the Web. Google Chrome, Firefox, Safari, and Internet Explorer all have extensions that will do their best to block advertisements. Though they're not perfect—and they're very much an arms race against advertising-based content providers—they work well enough, and the overall reduction in exposure to advertising is probably good for your head *and* your bank account.

Lastly, let's take care of your inbox. While, yes, you probably get a lot of important email, you probably get a lot of email that's not important too. Software can fairly easily tell the differences between these two things, and save the stuff that's not important for consumption later on.

Google has a tool for this in Gmail called Priority Inbox, but my personal favorite is called *Sanebox.com*. It works on most major email providers, and doesn't just mark what email is important—it actually takes the email that's not important, and dismisses it from your inbox into another folder. This way, the temptation isn't even there. Don't worry about missing anything. Close to the end of every working day, you'll get an email digest of all the emails Sanebox put into your "Later" folder so that you can go back and check to see what you missed.

I remember when I did this for myself the first time. The only thing I can liken it to is the first time I put on a pair of prescription eyeglasses. I didn't know my eyes and brain were straining to see things, but once the glasses were on, I could feel half the muscles in my face relax. It was a wondrous moment. It felt clean.

Sadly, it doesn't last long. After about five minutes, curiosity will kick in, and you'll start wondering what's going on in the world. You might start to panic. "What if there's an important email, or even a not so important but very banal and uninteresting email that's waiting for you in your inbox?" your inner voice might say.

Let this inner voice yammer all it wants, and treat it as though a crazy person has been locked inside your head. After a few days of working with Sanebox, I found myself questioning my very own significance: I was getting so much less email that I began to assume people didn't like me anymore. I found myself trawling through my Sanebox unimportant emails hoping that I'd missed some important email. My ego was wounded based on the sheer reduction in volume of email.

Now perhaps your ego isn't as fragile as mine, but I think that's the same voice that tells coke addicts to do more coke, and smokers to smoke more cigarettes. It's at best the voice of compulsion and at worst the voice of addiction. You're going to have to stop listening to it. It's going to take hard work, and a lot of strength, but you can do it. You just have to be pragmatic about it and take it slow.

Training

Watching the juggler jog backwards past me on the 16th mile, or the man towing his fully grown but clearly incapacitated son behind him on the 24th, was enough for me to understand that I probably wasn't going to ever be a world-class athlete. Though my wife looks like Indiana Jones trying to escape the giant boulder of "Clay"[81] behind her in our finish line photo, I'm still happy to have accomplished the feat.

I recommend that everyone in the world train for at least one marathon in their lives: it's a testament to what your body can do if you train appropriately. It's also a reminder that training takes a long time—training for that 26.2-mile race took six months—and a lot of small, slow steps to get there. For those of us that aren't world-class runners, marathon training is, above all else, a test of will.

To train for my first marathon, I used Jeff Galloway's "run-walk" method: it meant running for a certain length of time, then taking a shorter walk break. Over time, the goal is to lengthen the running time, and shorten the break time.

This method works for running for a few reasons. It allows you to go at your own pace and acknowledges that some people don't train for marathons in order to win them, but rather to complete them. It allows you to expend your energy based on your level of exertion, rather than on distance. Finally, it creates a framework that is growable and tunable to you as you grow.

We're going to use a similar method for strengthening your attention span. In order to do it, you're first going to need a good timer. You can buy a runner's stopwatch that has an interval timer if you want, or—if you want to be less annoying at work—you can download some software to do it on your desktop. Any old desktop timer will work, but interval timers work best. This will be the one notification you're allowed to have.

Before you really get started with this technique, I want you to promise to be forgiving with yourself. Failure at this doesn't mean you're a bad person, or that somehow you're not competent—it probably just means you bit off

---

81 I am so sorry for that pun. Remember, a healthy information diet is about having a sense of humor!

more than you could chew. Take it slow and find a pace for you. The thing that got my 240-pound self (after dropping 40 pounds during training) across a marathon finish line was this thought:

"I'm not going to win this marathon. My goal is to not come in last."

Now, we're going to start off slow. Try working in five minute intervals, with a one minute break in which you can do anything—check Facebook, deal with Twitter, or check your phone for text messages—anything you want, except check your email (we'll get to that in a minute). In one hour, try working like this five times, then pause your timer. Get up and stretch or use the restroom for one to two minutes. Remember, sitting kills you.

Once back at your desk, do another three repetitions of the cycle. By this time, you should have about 10 more minutes left in your hour. Check your email, and respond to the things that you need to respond to.

Now sometimes this won't work for you—you may *want* to pay more attention for longer spurts of time. That's fine; this is a framework, not a set of laws. These rules needn't apply all the time. But I will caution you— you're training for endurance, not short bursts of speed. Usain Bolt might be the fastest man in the world at the 100-yard dash, but it's unlikely that will do him much good in a marathon.

It's likely your mind will beg for you to work on a problem for longer than five minutes. In some cases it might be right, but stick with the program if you can. Even experienced marathon runners often run less distance than they can, so that they can train up for speed and better endurance; similarly, we're starting off at five minutes to make it easy on you—you need to get used to this pattern of working more than anything else.

So if you're working on complex problems, and feel that you must work longer than five-minute intervals, initially, then do it. But for a few hours, or even a solid day, give the 5:1 setting a shot. You might find that you get more minutes out of your day in the long term that way. Remember, we're starting off easy so that you don't get discouraged.

After you've got the 5:1 thing down, it's time to start increasing your attention span. In your first week, gradually turn up the number of work minutes in 15-second intervals. By day four, try to get up to about seven minutes. Remember to split your intervals up—in any given 60-minute set, you're going to need at least 2 minutes to stretch and about 10 minutes to deal with email.

As not all numbers divide into 48 evenly, these stretch moments and email moments are going to need pliable time limits. Do what you think is best, but if you have to err, err to the side of rest, not to the side of work.

By day 10, try for a 10-minute work interval to a 2-minute rest interval. A 10:2 interval may seem vastly inferior to a 9:1 interval. It's more than 60% less efficient to spend two minutes resting for every minute working as it is one for every nine. But remember, what we're trying to do here is to lengthen your attention span. At 10 minutes, we begin to get to the usual standard of our attention span length.

Continue growing your work time as you see fit, at increments that are shorter than noticeable. Do only 15- to 30-second increases, never more than once a day, and try not to go longer than 15 minutes without a small stretch break, at least. Remember: we're building a healthy lifestyle for you.

For this book, I worked in 15-minute work intervals with 2-minute breaks three times an hour, and a 9-minute email check at the end of every hour. I stretched, used the restroom, or otherwise didn't look at the screen for the full two minutes, I found this helped my mind reflect and decompress, so that I could get back to writing. Sometimes those two-minute breaks turned into five-minute breaks; sometimes those 15-minute work spans turned into 20-minute ones—I'm not a stickler for time anymore.

I also did only four hours in a row of this focused task work at a time, followed by at least an hour break that was entirely away from the computer screen. I tried to schedule my day so that I accomplished all the task-oriented computer work I needed to accomplish by noon, then I could take an hour for lunch. If I had meetings in any given day, they were scheduled for after lunch and if at all possible, back-to-back and directly after lunch. If my schedule allowed, then I was back at it after the meetings were over.

It's worth noting that I've started to apply this same technique to other things that require my concentration. Reading on the iPad, for instance, is tough for me because my email is just a tap or two away. In order to make a successful journey through a book, even for leisure, I've got to apply the same technique. The technique is about focus and concentration, not necessarily about getting work done.

The other good thing about this method is that it forces us to consciously measure the time we spend working on a computer. By building in the interval metronome, we become keenly aware of how much time has gone by, and how much time we have left to get done what we need to get done. No longer will you look up and wonder where the day went. You've used your executive function and accounted for it.

Finally, remember that you're measuring your success. We set up RescueTime for a reason: to make sure that what you were doing works for you. Make sure, after a week or two of doing this, that your productivity number is headed in the right direction, and that it stays that way.

All our brains and minds are unique, and though this works for me, it may not work for you. If it's the case that this system isn't working for you, then it's an opportunity for creativity. I encourage you to invent your own system for training out your attention span—and share it with us on *InformationDiet.com*.

Distractibility Can Be Good

It turns out that constant focus isn't all that great, and that allowing a bit of distractibility into our lifestyles can have some benefit. Several academic studies now show that surfing the Web mindlessly, for brief periods of time, can have restorative cognitive properties[82]—much more so than things with a high cognitive load like managing email. Focus on building your attention span, but don't forget to give yourself some breaks. Just make sure they're set to certain limits. Spending all day focused entirely on your work is bound to be exhausting.

---

82 *http://www.aomonline.org/aom.asp?ID=251&page_ID=224&pr_id=448*

CHAPTER 9

A Healthy Sense of Humor

"A merry heart doeth good like a medicine,
but a broken spirit drieth the bones."

—Proverbs 17:22

We could all stand to be a little more like Karl Rove in Washington. I met him once at a politics and technology event here in Washington, D.C. He glanced down at my attendee badge, saw the company name on it, and exclaimed: "Blue State Digital! You guys do great things for the wrong people!"

I responded: "Another half-truth from Karl Rove."

He laughed, told me how he would have beaten Howard Dean in 2004, and asked for my business card.

Three days later, I walked into my office to find a handwritten letter with a lot of strange stamps on it. It was a letter from my cocktail companion and renowned philatelist. The letter said:

"Dear Mr. Johnson,

It was a pleasure meeting you at the Yahoo! Citizens 2.0 Conference. Best of luck with the business, but only up to a point!

If you'd like to have the picture you took of me inscribed, please send it over and I'll sign it for you and send it back. If not, please accept this letter as a souvenir. Now you can show your liberal friends that you met the great Satan himself.

Sincerely,

Karl Rove"

Over the course of a few weeks, I found myself developing a pen pal in Rove. He and I exchanged a few letters—he romanticized "pushing atoms back and forth"—and I thanked him for helping us raise all that money for

MoveOn.org. (Though I must admit, my interface with the United States Postal Service isn't what it ought to be.) It struck me that Rove, arguably one of the most successful political architects in history, was not only funny, but he was also keenly aware and capable of poking fun at himself.

Left-of-center people may find this atrocious. Here's a man who helped architect George W. Bush's political strategy for eight years. Known as "Bush's Brain," he's often thought of by millions of people as one of the most evil spinmasters that ever existed—alongside Dick Cheney, and Bush himself, Rove is thought of by the left as the puppeteer behind the administration that led us through the Iraq war, the botched Katrina efforts, the housing bubble, and the banking system's meltdown.

And here he was, in a courteous and handwritten note, being hilarious though downright glib about the whole thing. In Rove's defense—he wasn't making fun of those things that the left has him tied up in. He was making fun of himself.

Rove has a sense of humor because he has to, and he probably understands the same thing I've learned in the past few years of working on issues that I deeply care about and things that appear, to me at least, to be vital to our condition.

Chances are, if we can't laugh at something, we can't think rationally about it. (The exception to this rule is sports. Sports is about clear wins and losses, and most importantly, entertainment. It's okay to polarize sports— it's not any fun if you don't. The last thing we want to do is think rationally about sports. The stuff that matters, though, is about our livelihoods and the future of our country.)

Laughter is important to a healthy information diet because it has all kinds of incredible health benefits. It turns out laughter increases our heart rate in a good way, increases our cardiovascular health, and burns calories. Some science shows that laughter may cause increased blood flow to the brain and decrease stress (thus boosting our immune systems), may normalize blood sugar levels, and may help us sleep better.

The first way a sense of humor helps is that it makes the truth more palatable. It bypasses our gut reaction for fight and flight, and makes it comfortable to hear what's going on in a more digestible fashion. Shows like the Daily Show and the Colbert Report help us to find humor in the daily news, but they also tend to feed us small nuggets of truth wrapped in delicious, bacon-like hilariousness. Sometimes, it's a healthy way of getting some national news exposure without having to take stuff too seriously.

But watching the Daily Show isn't going to give you a sense of humor, and relying on it solely for your national news information diet is likely to leave you with a point of view that's just as misinformed as watching FOX or MSNBC. Jon Stewart himself will tell you: his job is to entertain you, not to inform you or even tell you the truth. The difference between Jon Stewart and Bill O'Reilly is that Stewart is honest about his role as entertainer.

While these shows are funny, watching them isn't the same as having a sense of humor. We shouldn't conflate laughter with having a sense of humor. Laughing is important, sure, but being able to see the humor in all things—especially yourself—is even more important.

It turns out that a sense of humor might just be a vital part of our brain's ability to rewire itself.

Much of what makes us laugh are things that are unexpected. The great jokes are about misdirection and surprise. As we anticipate the punchline of a joke, we're trying to figure out where it's going—the joke itself tends to be a buildup towards an expectation, and then comes the punchline: usually something unexpected. That's what makes it funny.

Take Rove's letter: he leads with something rather standard—a greeting and formality, but then closes with a killer punchline. It immediately changed my opinion of Rove, unwiring the heuristic in my brain that's been trained by years of being a democratic political operative to believe that the man is pure, unbridled evil.

Instantly, upon reading that letter, Rove became to me somebody that's human and very aware of himself. My presumptions about him changed and all of a sudden, I found myself saying in my social circles, "Oh, Karl Rove isn't so bad. He just has different beliefs than we do." I'd get jumped on, then pull out the letter and show it off. The power of Karl Rove's humor softened the hearts of even the most liberal of activists.

It turns out that there may be some science behind this idea. In their book *Inside Jokes: Using Humor to Reverse-Engineer the Mind* (MIT Press), scientists Matthew M. Hurley, Daniel C. Dennett, and Reginald B. Adams Jr. provide a cognitive and evolutionary perspective for our sense of humor.

They argue that humor could be a cognitive cleanup mechanism of the mind, that nature needed a way for us to constantly check our judgmental heuristics, and reward ourselves for seeking the unexpected. They stipulate that laughter itself is a social signal that demonstrates cognitive prowess— something that's useful in mate selection—and thus, our ability to laugh spread through generations.

Humor tends to be a useful mechanism for figuring out when you're overly attached to information, too. If you can't laugh at something, it likely means you're not flexible with the information—that you take it so seriously that your mind cannot be changed. While it's good to have these stances on some topics (say, the Holocaust or slavery), if you can't laugh at Lebron James jokes, you might be taking your love of the Miami Heat a little too seriously.

Studying humor tends to make whatever might be funny no longer so, so I'll leave it at this: lighten up.

CHAPTER 10

How to Consume

"While it is true that many people simply can't afford to pay more for food, either in money or time or both, many more of us can. After all, just in the last decade or two we've somehow found the time in the day to spend several hours on the Internet and the money in the budget not only to pay for broadband service, but to cover a second phone bill and a new monthly bill for television, formerly free. For the majority of Americans, spending more for better food is less a matter of ability than priority."

—Michael Pollan

So now we've got our three skills: data literacy, a sense of humor, and a method for training and accounting for our executive function and attention span. The question now is: what is it that we should consume? What kinds of information go into a healthy information diet?

The world of food is littered with advice, and the one we probably know the best comes from the United States Department of Agriculture: the food pyramid. You've seen it—it looks like Figure 10-1.

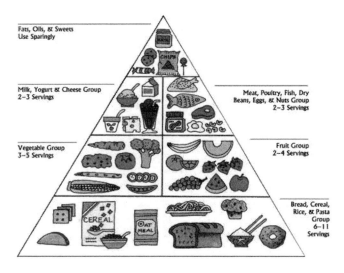

Figure 10-1. The United States Food Pyramid: 1992–2011.

In 2011, the food pyramid was found to be too complicated, so it was distilled into something a bit more simple, *ChooseMyPlate.gov*, shown in Figure 10-2.

Figure 10-2. The New ChooseMyPlate.gov: 2011–.

There is currently no government agency to monitor information consumption—though former President Bill Clinton suggested creating one in May 2011.[83] He suggested an agency that would regulate our information providers and suggest to us what information we should consume and which we should not—an independent agency run by the government that would determine what kinds of information ought to be released.

---

83 *http://www.politico.com/news/stories/0511/54951.html*

I suspect that if this idea gained any serious traction in government, the public would loudly destroy the thought. It's just not viable: the first amendment prohibits any authority a federal agency could have over speech, and even if that were miraculously overlooked, it'd be a waste of money. The agency would have zero credibility with any consumer of information.

It'd be immediately labeled an Orwellian "Ministry of Truth." We can regulate food because neither beef nor turnip greens typically inform our vote. Moreover, it'd be impossible to label and classify all the kinds of information we consume. A nutritional label (see Figure 10-3) would be equally impossible and ridiculous.

But just because there shouldn't be a ministry of information, it doesn't mean there shouldn't be "dietary guidelines" for information. They just shouldn't come from government. Ideally, they ought to come from science, however you won't find many neuroscientists clamoring to build the info-pyramid.

And of course, with food, we have the aforementioned 60,000 diet books available on *Amazon.com* if we want to get more advanced. Unfortunately, we don't have the same kinds of resources to draw upon for creating a healthy information diet book, since it's difficult to dissect individual information resources for their exact nutritional values. On top of that, proposing a list of information one should take in seems nearly reprehensible—who am I to tell you exactly what you should be reading?

Nutrition Facts

Amount Per Serving	
Time: 30 min	Opinion Time: 18

Total Opinion	
Partisan Rhetoric	36%
Expert Analysis	8%
Advertising	22%
Cited Sources	2%
Historical Favortism	

Ingredients: Paul Krugman, Ann Coulter, U.S. Bureau of Labor and Statistics, Toxic Waste Inventory, H.R. 1234

Figure 10-3. A sample information nutrition label.

I don't want to tell you *what* information to consume, or impose my own biases on you—that wouldn't be responsible. Instead, I want to give you a framework for information consumption. Like nutrition, you won't be nearly as successful at this if we focus too much on the food itself; instead, we have to focus on developing healthy habits for information consumption.

But, as I've noted before, the Internet moves faster than an author or a publisher, so if you want the latest and greatest resources, please visit *InformationDiet.com* and visit the wiki where I, along with the community of other readers, will keep an updated list of reliable sources at the bottom of the trophic pyramid.

Consume Consciously

Let's first define the kind of information consumption that matters for our discussion. When I say consumption, I mean the kind of consumption that requires action on your part to initiate, with something whose purpose it is purely to provide you with information. Watching television, surfing the Web, listening to the radio, playing video games, and reading books, magazines, or newspapers—these are all forms of active information consumption. If it has a channel, a page, a frequency—if it involves you turning it on and off, or you picking it up—that's the kind of information we're talking about.

We're *not* talking about the information consumption you don't have explicit control over beginning and ending: advertisements on the side of the road during your commute to work, conversations with friends, families, and the waiter at your local restaurant, or the music in an elevator. While these things do contribute to your overall information intake, you don't have a lot of control over them, so we can't do much about them without turning you into a recluse.

We're also not talking about the production of information for others to consume. While this is part of data literacy as we discussed earlier, writing, outlining, and even editing shouldn't count towards our total information consumption.

Keeping It Clean

You'd never be successful on a food diet if your freezer was filled with ice cream, your refrigerator was filled with fried chicken, and your cabinets were filled with macaroni and cheese. So first let's clean out our metaphorical information refrigerator.

I advocate canceling your cable or satellite television subscription if you have one, and getting your video entertainment from services like YouTube,

Hulu, and Netflix. With the exception of weather information, most news services carried by television networks don't do the public any service. Having cable (or satellite TV) in your home while being on an information diet is like trying to go on a food diet with a magical sink that pours not only hot and cold water, but also delicious milkshakes. While you may have the will to resist it, let's do what we can to increase our chances of success.

This move is also economical. A basic cable package, on the lowest end, costs consumers an average of $52 a month or $624 per year. Add in premium stations and advanced packages, and you'll see your cable television bill approach upwards of $100 a month or $1200 a year.

With a reasonable broadband connection, even if you purchase individual episodes of television at $2 an episode from a service like iTunes, you end up with a net annual savings, and many other benefits, including not having to watch advertisements, resulting in saved time. You'll also remove the temptation to couch surf and mindlessly watch any show being provided to you.

But besides saving you money, cutting cable is going to start changing your relationship with information—and shift you from being a reactive consumer to a conscious one. If every piece of information you consume on your couch comes with a cost, or at least involves more conscious selection than flipping through the long list of what's available on cable at a given time, you'll have more control over what you're consuming.

Tim Ferriss, in his book *The 4-Hour Work Week* (Crown Archetype), advocates for an information diet that he calls *selective ignorance*. It first involves fasting: not checking email, not dealing with social networks, and avoiding much of the "incoming" information you have for a solid period of time. During this time, one allows only a deliberate intake of one hour of non-news information on television, and one hour of fiction reading per day. Then you wean yourself back onto an information diet of only information that's actionable and relevant.

For most, I think this will yield an unsuccessful outcome. By the end of the fast, you'll be so eager to plug back in that—like a food fast—you're likely to binge as soon as you get the chance. The selective ignorance plan also encourages us to eliminate diversity in our information diets, rather than exposing us to a diversity of knowledge, information, and opinion that may come our way.

I prefer a data-driven and more pragmatic approach. When you start a food diet, the most sensible way to figure things out is to first audit the calories you're taking in, to see if you're overconsuming. An honest food journal can help you keep your food intake under control.

We should try the same approach with information. We need a framework for figuring out how much information we're consuming if we're to consume more of the good stuff and less of the bad stuff. There are three ways we can measure our information intake: by the number of words we hear and read every day, by the amount of overall bytes we consume (like we measure computer intake), and by hours—the amount of time we spend deliberately consuming information. There will always be more bytes and more words, but time is non-renewable, so let's use this as our method of measurement.

Take a liberal count of the hours you spend in front of a computer consuming information for one week. You can do this in two ways: by keeping a journal and spending two minutes at the top of each hour estimating how much of your time you spent consuming—or automatically, by using a time-auditing tool like RescueTime, and then estimating the amount of noncomputer time you spend afterwards.

Since we're using time as our measurement, it makes sense to use scheduling as our form of information intake. If we stick to a schedule, we're exercising control over it, rather than allowing it to control us. It will also help us to respect our information-intake time. By allowing ourselves only a finite amount of time in which to consume information, we can consume more deliberately.

I recommend trying to slowly adjust to an information consumption time of no more than six hours per day. For some of us—the knowledge worker especially—this sounds impossible. But look at it this way: the professional's job is to produce, and if you're spending less than half of your work day on the production of information, you're likely not being as productive as you could be.

A sample information intake schedule may look something like this:

7 a.m.–8 a.m.: Information consumption time. Read the newspaper, watch morning television, check the weather, check social media feeds, etc.

11:30 a.m.–12:30 p.m.: Email

4:30 p.m.–5:30 p.m.: Email

8 p.m.–10 p.m.: Entertainment time. Watch television, check social media feeds, etc.

10 p.m.–11 p.m.: Fiction reading

For the person with Gmail or Outlook living permanently on her desktop, Twitter scrolling by in the background, and Skype and Google Talk running in the background, even the idea of this schedule may cause heart palpitations. It's a strict, low information schedule involving only two hours of email, four hours for entertainment, and zero hours for education or research.

This schedule is a framework of what your information diet *could* look like, but it's not written in stone. Some days, when you need to do a lot of research or you feel the urge to learn something new, you might move things around and consume less entertainment or less email than you would on another day. Some days, your information diet will just require you to consume more information than others.

The important part isn't what you spend your time on or when you spend it. The important part is that you create a flexible schedule for yourself and stick to it.

For the average person, who currently consumes more than 11 hours of information a day, I do not recommend jumping straight into the six-hour information diet. Instead, try to wean yourself slowly. Give yourself achievable goals. Audit the time you're presently spending consuming, and start reducing it by 30 minutes every week until you get to a time that's right for you, your goals, and your job.

For many, this will result in a net increase in our most non-renewable resource: time. A six-hour consumption day is truly terrifying for some, not because they're afraid of no longer being connected, but because they won't know what to do with the extra time. If you're cutting five hours off your information intake time, you're going to need to divert your attention to something else during those remaining hours.

Try to fill some of those reclaimed hours producing, rather than consuming, information. Try writing in a paper journal, writing articles for a blog, taking up photography, or creating funny videos of kittens for the YouTube audience, if you must. As we discussed in Chapter 7, Data Literacy, the production of information sharpens the mind and clarifies your thought.

You can also increase your social time, spending time talking with your spouse, family, and friends. Another good use of your time is giving your mind a chance to digest the things that you've read by taking long walks, spending time exercising, or even meditating.

Nutrition isn't just about what or how much to eat, it's about eating balanced meals. Just like the new recommendation graphic from the government recommends that our plate consist of 30% grains, 30% vegetables, 20% protein, and 20% fruit washed down with a glass of milk, we've got to come up with a healthy means of consciously consuming information.

Unfortunately, we can't make an exact replica of *MyPlate.gov* for information—we don't have the kinds of neurological research out there to figure out what a healthy, complete diet truly looks like. But like Banting, we do know the kinds of things we ought to consume less of.

Mass affirmation is the refined sugar of the mind—I'm not talking about the kind of relatively rare positive affirmation you get from friends or family, telling you that you're loved and respected. Rather, it's the mass affirmation: the affirmations you get that aren't intended for you specifically, the stuff that television is best at, but also permeates through all of our information delivery mechanisms. The suppliers that make a living telling you how right you are are the ones you ought to avoid the most.

I try to limit myself to no more than 30 minutes a day of mass affirmation, and strive to consume much less. It means making some tough choices, and letting go of some things you might enjoy. At a maximum of a half-hour a day, for some liberals, it means having to make the dreaded decision of choosing between Stephen Colbert, Jon Stewart, and a regimen of DailyKos. For the conservative, it may mean having to pick between Fox and Friends for a half-hour in the morning, and a half-hour of Bill O'Reilly in the evening.

Consume Locally

The Hitchhikers Guide to the Galaxy (Del Rey) by Douglas Adams starts off with the book's protagonist, Arthur Dent, having his flat completely demolished by a local government agency, thanks to an eminent domain ruling to build a highway through his apartment building. The notice, it's written, was to be found in the bowels of a government building, up for "public display."

Perhaps Dent was too obsessed with *US Weekly*, or news from far away. Our obsession with national news over local news has to end. While it's important to stay abreast of national and world affairs, most of us give too much weight to information that's not actionable and relevant to our daily lives. There are more dealers of junk, more profits involved, and more lies to be told as we sit higher on the trophic pyramid.

A healthy information diet means the avoidance of overprocessed information. A healthy information dieter constantly tries to remove these junk dealers from the consumption chain. That means either consuming locally or working consistently to remove distance to the things that you investigate.

Consuming low on the metaphorical trophic information pyramid doesn't mean just sticking closer to the facts; it also means that it's easier to stick close to the facts when you stick close to home. The further away from home you get, the more attention you have to pay to how many operators have been involved in getting you that information.

Facebook's founder, Mark Zuckerberg, once quipped, to the alarm of many an activist: "A squirrel dying in front of your house may be more relevant to your interests right now than people dying in Africa,"[84] and while I too bemoan the trivialization of famine, genocide, and HIV—he has a point. You alone can go get the squirrel and clean it up and prevent your neighborhood from smelling like dead animal. Chances are, you're unable to solve famine, rid the continent of Africa from evil warlords, and cure HIV all by yourself. Local news is more actionable and relevant to the individual than global issues.

Luckily, there's a renaissance going on in the world of local news—new tools allow you to get online and see news and information down to the narrowest geographic criteria possible: your block. Today, major cities and government agencies are releasing information by the gigabyte that informs us on the real goings-on in our neighborhoods.

If you're in one of the dozens of cities lucky enough to be covered by Everyblock, I highly recommend it as an important daily source of information. The site aggregates dozens of data feeds that come from local governments and turns them into an easy-to-read, relatively opinion-free way of seeing what's going on at the block level—and you'd be surprised how much information there is about your single block.

Everything from bulk trash pickups to police reports to photos taken in your neighborhood to recent real estate listings are available for you. You register for the service, plug in your address, and tell the service whether you are interested in getting information about your city, your neighborhood, or the area within an eight-block, four-block or even one-block radius of where you live.

The site also allows you to post messages to other people in your neighborhood so you can talk about the issues affecting your real, local community. It makes the information that comes out of your community immediately actionable, and allows people to connect with their neighbors easily.

---

84 http://books.google.com/books?id=PxTvbM-VCPEC&lpg=PA296&ots=DSf8nQQX5i&dq=a%20squirrel%20dying%20in%20front%20of%20your%20house%20may%20be%20more%20relevant%20to%20your%20interests%20right%20now%20than%20people%20dying%20in%20Africa%20Zuckerberg&pg=PA296#v=onepage&q=a%20squirrel%20dying%20in%20front%20of%20your%20house%20may%20be%20more%20relevant%20to%20your%20interests%20right%20now%20than%20people%20dying%20in%20Africa%20Zuckerberg&f=false

Beyond Everyblock, your city may have its own data catalog available for you to peruse. Most major cities—like D.C., Portland, Seattle, San Francisco, and New York—have them, and more are on the way. To find yours, do a Google search for your city's name and the phrase "data catalog" and you'll likely stumble upon it. If you cannot find one, try searching for the email address of your local government's CIO (Chief Information Officer), writing to them, and asking them to make the data feeds that they have available online.

Public data is your data—you fund its collection with your tax dollars, and it ought to be taken out of the silos in the underbellies of city halls across the country, and into the light of day. Most sunshine laws—laws that require governments to respond to citizen requests to be open—require government officials to respond to citizen requests for information. Be an activist, and ask for its release.

It's hard to be factually incorrect about this kind of data, and reading about parking in your neighborhood may seem quite dull. Over time though, you're able to spot trends: observing a string of car thefts on your block may yield you some pertinent information—certainly more pertinent to your safety than whether the federal government is going to invest in high-speed rail.

For news, reading your local paper, watching your local news when it's on, or reading local blogs isn't a bad idea, but keep in mind: you're now becoming a secondary or tertiary consumer of information, and you're more subject to succumbing to your own bias and other forms of misreporting.

While this information is less likely to be as manufactured as what you'll find in the national and international news, it will still require some work in order to make sure it's trustworthy and verifiable. In order to consume this information safely, you must do the extra work of investigating source material, figuring out the intent of the person delivering that information to you, and determining that information's effects on you.

The local news renaissance is also a renaissance in specialized, deep wells of information. Instead of grazing on global and national news, and information about people you don't know and who don't care about you, shift your information consumption to local news and people who do care about you. Try to achieve deeper relationships with the information you're consuming: if you must consume information about the affairs of people and places far away, try slicing off a niche, and developing a mastery of it.

But geographically local information isn't the only kind of local information we can get to. Socially proximate information also sits near the bottom of our informational trophic pyramid. Like geographically local information, socially local information—information about the people closest to us—is actionable, relevant, and important to our connections with other human beings.[85]

The Web gives us new ways to check in on those we know and love, even when they're far away. But like all other forms of information, social media comes with consequences. We have to filter the information that our friends are sharing about themselves and the information that they're resharing from elsewhere.

It's good to fine-tune your lists of friends and acquaintances and fortunately, all of the major social networks give us this ability. Facebook's groups and lists, Google+'s circles, and Twitter's list functionalities make it so that we can sort our friends and view our social networks through the lenses of what's important.

If you are a user of one or more of these services, take an hour or two and sort through your lists of friends. Create a group, list, or circle for family members, another for close friends, another for work colleagues, and another for people you'd like to get to know better, and read those posts consciously during set periods of the day, rather than plunging yourself into an ever-growing stream of incoming media that your brain will be unable to resist.

Low-Ad

We've adjusted our information culture such that we now expect information to be free to the consumer. But that free information comes with a much higher cost: advertising. A healthy information diet contains as few advertisements as possible. The economics of advertisement-based media make it so that our content producers must draw eyeballs in on every piece of content, and that results in sensationalism.

Sensationalizing content tends to degrade its quality. That's not the only cost, though: because advertising persuades us, over time, to buy things that we wouldn't ordinarily buy, the cost of consuming ad-supported content is higher than we think. I know I've ordered a pizza or two from my local pizza joint after watching a television commercial for Pizza Hut.

The reality is that so much of our information—even information we pay for—comes along with advertisements, and it's nearly impossible to escape advertising completely. Our routes to work and our walks down the street are filled with advertising, and even if we manage to escape those, our trusty letter-carrier delivers more directly to our homes for us to see.

---

85 If you still have one, call your mother.

Part of a healthy information diet is respect for good content, and a disrespect for advertisements. We have to reward our honest, nutritious content providers with financial success if we're going to make significant changes. I subscribe to *ConsumerReports.org* and *NationalGeographic. com* as a paying member because they provide good, high-quality, and mostly ad-free content to their subscribers.

A healthy information dieter most certainly won't sign up to receive advertisements—though many of us do. Our email boxes are filling up not just with spam, but with the latest travel deals from Expedia and specials from JC Penny and *Amazon.com*. Unsubscribe from these lists, or create a filter or rule in your email client to remove them from your inbox.

While it's likely impossible to be informed and ad-free, it ought to be something to strive for. To limit your exposure to advertising alongside content, I recommend using tools like *Readability.com*. Readability gives you the ability to remove distraction from content—it removes advertising completely from any article you're reading, gives you a more readable typeface, and adjusts the width of each article to make it easier to read.

Readability incorporates another application called Instapaper in its service. It is a similar tool that also allows you click a button on your web browser and move the article to your mobile device. With Instapaper, you can find articles you'd like to read, and read them more easily and more free from the distractions of advertisements and suggested reading headlines on your iPad, iPhone, or Android device, or through Instapaper's service on the Web.

Knowing that they're circumventing the current advertising distribution model of information, Readability charges a minimum membership fee of $5.00 per month that you can increase to however much you want. It takes 30% of the membership fee as its own, then allocates the remaining 70% to the content providers that you read through the service. It's an invisible, transparent way to support content providers without having to wade through advertisements.

The websites of all content providers are designed to keep you reading, and to expose you to the most advertising impressions possible. It's why they split articles up into several pages, and why when you scroll down to the end of an article, you're plied with more enticing articles to read.

Instapaper and Readability help to reduce your exposure to these time-sucks, and help you retain a sense of conscious consumption. The key part of these tools is that they make it easier for you to focus on what it is that you want to focus on, and eliminate the distractions you'd normally encounter. They make conscious consumption easy—instead of blindly surfing the Web and reacting to what's being thrown at you, you can instead shop for content, select the things you want to read, and then have a longer reading session free from distraction.

Diversity

Processed information isn't the only thing to avoid. If we are comparing an information diet to a food diet, then affirmation of what you already believe is the mind's sugar. A healthy information diet means seeking out diversity, both in topic area and in perspective.

A healthy information diet means affirming our beliefs only to an extent, keeping a watchful eye on our own fanaticism, and soaking up as much challenge to our beliefs as we possibly can. Getting perspectives that agree with you is one thing, but getting *only* perspectives that agree with you is bad for you—it may limit your exposure to good information and may cause you to suffer from the forms of ignorance I described earlier. Moreover, it's through having your ideas challenged (and through the synthesis, analysis, and reflection of those challenges) that your ideas get better.

Fried chicken and ice cream are okay to eat every once in a while—at most, a few times a year when you're celebrating or feeling particularly down and just need some comfort food. The same goes for the news sources that provide you with the most comfort and information, or even antagonize you. Recognize them as primarily entertainment, and treat them like rare, special servings rather than as something representative of your daily intake.

Striving for synthesis is necessary, and that means actively encouraging a diversity of opinion at all levels of your information diet. Remember the story of Eli Pariser and the filter bubble: we never want a personalization algorithm to start thinking that we're only interested in hearing viewpoints from one particular side, one particular class of people, or one particular topic or issue.

Asynchronous social networks (ones where you can follow someone without them following you back) like Twitter and Google+ allow you to craft a diverse set of information inputs. You can choose to balance your inputs by following people with a different background or point of view than yourself and your closest friends to get a better perspective, or to learn where people who are different than you are coming from.

Without constant attention to perspective diversity, we assure ourselves mutual intellectual sycophanticide. Because human beings tend to self-select into self-reinforcing groups, tools like Facebook and Twitter allow us to get not only constant updates from our friends, but also constant affirmations of our beliefs. Only through constant pruning, selection, and conscious clicking can we make them work for us.

In other words, the only thing to be fundamentally opposed to is fundamentalism itself. To help counter this, I keep a bias journal on my computer, but you could just as easily have it written down on paper if you like. In it, I keep my firm positions and values—stuff I find to be absolute. It's just a simple, noncategorized list of strong biases I may have. Here are some of mine:

- Affordable access to quality healthcare is a fundamental right

- Innovation in the private sector will always outperform innovation in government

- Large organizations are less interested in the individual than small ones

- Strong affinity for Google products (could be because I get invited to speak at their conferences)

- Strong affinity towards technical solutions for social problems

- Men who wear brightly colored Pumas are annoying

Some biases are stronger than others, of course, but what's important is that you're honest with yourself about what your biases are. Some of them could be deeply private, but you don't have to share your list. What's important is that you keep the list, are explicit about it, and constantly look to find data and people that challenge your biases—and prescribe yourself enough time to encounter them.

It's also important to seek out diverse topics of information, as the synthesis of information from different fields helps us create better ideas. It also helps keep us from losing our social breadth—so we have more to talk about than the specialized knowledge of our particular fields. Introduce some new ones into your information diet. I find three resources particularly useful in this regard.

The first is the Khan Academy. Started by Salman Khan in 2006 in order to tutor his young cousins, the site now features over 2,600 small lectures on anything from basic subjects like arithmetic and European history to advanced subjects like organic chemistry, the Paulson bailout, and the Geithner plan to solve the banking crisis.

Being an infovegan means acquiring the basic knowledge you need in order to understand what the data is telling you. The Khan Academy opens the door and lets you in. It's not a good stopping point, but it's an excellent way to pick up the basics of a subject that will give you the knowledge you need in order to conduct further research.

The second is TED (Technology, Education, and Design), an organization that puts on a conference every year. It invites luminaries from a myriad fields to come and present what they're working on, and then share the talks online via its website. TED talks—especially about things you're not ordinarily interested in—are a great way to add diversity to your diet.

The third is Kickstarter, which has effectively replaced the "Arts & Leisure" section of my local newspaper. Kickstarter's purpose is to fund small projects and help artists and entrepreneurs get off the ground, but it turns out that it's grown to be a good source of inspiration and entertainment as well.

Kickstarter lets you see what some people (the self-selecting group that uses the service) are passionate about—whether it's building the world's largest database, performing analysis of hip-hop music, or writing a guidebook to breakfast joints in Columbus, Ohio. It lets you browse local projects, too, so you can see what kinds of things are starting up in your town—and if you feel inclined, you can support local artists.

Again, the point isn't to visit these three sites as an endorsement of ideas, or a strict rule for your information diet. But in the frame of conscious consumption, they mean something different. You're choosing to consciously visit these sites on a regular basis in order to get something particular out of them: diversity.

Think of it like going out to a different kind of restaurant than the usual places you go. There's nothing wrong with eating at the same place every day, but sometimes you need to branch out and see what else is out there.

Balance

So just how much of what should you consume? Every diet book in the world has some kind of recommendation—an interesting way of telling you what it is you should eat, and what it is you shouldn't. I'm afraid that in the world of information, our tastes are far more diverse and require far more specialization than our food diets, and thus, I can't make a recommendation for everybody.

There's also information I've left out—information that I'll make no attempt to classify or prescribe a diet for. For instance, our varying religious beliefs have prescriptions for consumption that are inappropriate to contradict.

Our information diets are also tied to our professions. Nobody but models and personal trainers get fired for eating too much fried chicken, or promoted for eating too much celery. But our information diets have serious job consequences: a doctor not dedicating enough of her time to skill development could lose her ability to practice.

Because our jobs and belief systems are very different, and because professions and religions often come with their own basic information diets baked in, a universal prescription for an information diet is impossible. But the good habits I've described in this chapter are possible.

The information diet I maintain looks like Figure 11-4.

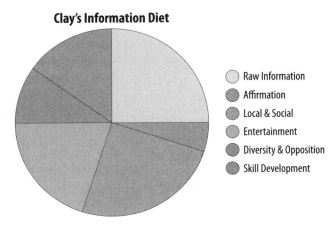

Clay's Information Diet

- Raw Information
- Affirmation
- Local & Social
- Entertainment
- Diversity & Opposition
- Skill Development

Figure 11-4. The information diet maintained by the author.

The categories I've chosen here reflect the various suggestions I have in this chapter, but your breakdown will look different than mine. The situations of your work, and your stage in life, may require a vastly different diet than the one I'm on. And the truth is, the averages I've suggested are averages: they vary from day to day. Pollan's "Eat. Not too much. Mostly plants." beats the food pyramid not only with its simplicity but also its flexibility.

The classification and categorization of information are always subjective, and sometimes controversial. Do not worry nearly as much about achieving some set standard of balance, or even emulating my diet. Worry about consuming consciously, and making information—and our information providers—work for you, rather than the other way around. Form healthy habits, and the right balance will follow from it.

Balance means keeping our desire for affirmation in check. For the amount of time I spend consuming things that I believe in, I try to spend twice as much time seeking information from sources that disagree with me. The end result is twofold: not only do I gain exposure to differing viewpoints, but I also limit my passive exposure to mass affirmation.

Support and Fine Tuning

Going on an information diet is as difficult as going on a food diet. For a lot of us, it requires the support and ideas of our family and community. And it's personal, too—our minds, just like our food palates, have different and unique tastes. Building a healthy information diet means discovering what works best for you, and creating a routine that you can stick to.

I built *InformationDiet.com* with this in mind. Reading this book is just the beginning of what is hopefully a larger journey towards better health, and as more people make more discoveries about what works for them, we can start sharing with one another what works and what doesn't.

If you're looking for ideas about what kinds of information could possibly share in your information production regime (I recommend at least an hour a day dedicated to writing or otherwise publishing information), try publishing what your information diet is, and how it's working for you. Publish it on the InformationDiet.com forums, or publish it on your own website and drop me a line on Twitter (@cjoh), and I'll be happy to link to it from *InformationDiet.com*.

Social Obesity

*"If a nation expects to be ignorant and free, in a state of civilization,
it expects what never was and never will be."*

—Thomas Jefferson to Charles Yancey, 1816[86]

When we start looking at information consumption through the lens of a diet and take responsibility for the information we're consuming, things start to get really frightening. Poor information diets and poor filters are responsible for really atrocious things and have horrible social effects that are, as history suggests, as deadly as the worst of our diseases.

Physical obesity, it turns out, may be a social contagion. Some studies suggest, for instance, that introducing an obese person into your social circle may put you at risk for obesity. It's not hard science, and there is disagreement—the counterargument to these studies is that we tend to homogenize in groups, so people who are already obese may just associate with one another, and reinforce one another's bad eating habits.[87]

Regardless of causality, this trend is something we recognize from common sense: hang out with people living healthy lifestyles, and chances are you'll be exposed to more stuff that's good for you and less stuff that's bad for you. If all your friends are alcoholics, it makes it more difficult for you to quit drinking. Because our consumption of food is tangentially social, those with whom we choose to associate affect our intake.

Information is far more social than food. You can grow your own food, and eat by yourself your entire life, and still remain healthy—but if you were the only person on the planet who knew how to speak, read, and write, you'd likely go crazy.

---

86 *The Writings of Thomas Jefferson*, Memorial Edition 14:384

87 *http://citeseerx.ist.psu.edu/viewdoc/download?doi=10.1.1.165.3862&rep=rep1&type=pdf*

Because information is social, information diets have far more severe social effects. Just ask Alfred Dreyfus.

In 1894, Alfred Dreyfus was a 35-year-old Jewish man, and an unknown captain in the French army, who was rushed to Court Marshall and life imprisonment for allegedly leaking French secrets to the Germans. The evidence against him? A crumpled note in a trash can with a single initial on it. Though his handwriting didn't match the note, he was accused of disguising his own handwriting, and exiled to Devil's Island.

The French were swept up in a divisive debate over the man's guilt. The debate gave birth to a new word, *intellectual*, which was not intended to be a compliment. Instead, it was a derogatory word for the people supporting Dreyfus' release—equated with someone too introspective to be loyal to one's military and one's nation.

The head of the intellectuals was French writer Emile Zola, who famously wrote, "J'Accuse...!" in an open letter to the President of the Republic, describing the nonsense of the case that would eventually become known as the Dreyfus Affair.

It took 12 years of bitter public fighting for those intellectuals to win, and more for the French to recover. Dreyfus and the French are not the only victims of fights like this. The genocides in Rwanda were fed by hate speech on the radio. Hitler's embrace of the new media of film empowered Nazism. Humanity's darkest moments are the ones in which masses of people had the worst information diets.

Today, we're fighting a million Dreyfus Affairs with one another. Rather than focusing on issues, we've tribalized into a million little rights and wrongs. In Washington, our completely polarized electorate is distracted from serious, solvable problems because those problems aren't salient or interesting enough for them to pay attention. What makes for good politics doesn't make for good democracy.

Why would someone pay attention to the major problems that we have with the federal acquisition regulation (which directs how government spends money on contractors) when we have to "win" on the debt ceiling vote? Why talk about measurable successes in our classrooms when we can have fights over the teachers' right to form a union?

You might argue that stupid people, willing to believe whatever they want to believe, will always exist. You might further argue that evildoers will always be there to attempt to take advantage of them. You're right. But the problem is getting more severe because the economics of how we get our information have changed so much that it's not just the stupid people who are getting duped anymore.

The only way we can solve the problem of information obesity is to change the economics of information. And while it's not going to solve itself overnight, it's an issue that, with enough demand from the consumer, will begin to change. Just look at what's happened with the healthy food and local food movements.

Welcome to the Vast Rational Conspiracy

Part of the reason people have poor food diets is that the food that's cheap tends to be the food that's the worst for us. Thus, there's a strong relationship between poverty and obesity in the United States; it turns out that our poorest counties are also our most obese. But there is a way to change that.

As a result of consumers demanding healthier food, and a public concern about obesity, Walmart is attempting to cut up to 25% of the salt, fat, and sugar from its foods in order to combat obesity. Because of demand, Walmart is now the single largest provider of local, organic foods to the market. The result: the entire food industry is changing and following suit so its foods can be sold in Walmart stores.

It's not just taxes and smoking bans on cigarettes that drive down the number of smokers in the United States. There's also social consequence to smoking. Now, smoking isn't just something that causes cancer: for many, it's something that's socially unacceptable—a cultural faux pas. The smokers have been dismissed to our back alleys, behind the buildings. More and more, they're forced to hide their habit, which in turn creates fewer smokers.

We can do the same with our information providers, but only if we show consumer demand for high quality, source- and fact-driven information. The market will move, but only if we show that there's a positive economic outcome from doing so. If we start to change our information consumption habits, the whole market will change to follow suit. If Fox and MSNBC are no longer rewarded for being affirmation distributers, and their ratings start to change as a result, it will have consequences not just for the information dieters, but also for the public en mass.

An information diet isn't just something that's good for you. An appropriate diet is a social cause that yields a better ecology of mind—one that's more immune to contempt and hate, and to the tragic consequences of what those emotions beget.

If we begin to demand an end to factory-farmed content, and instead demonstrate a willingness to pay for more content like investi journalism and a strong, independent public press, we'll not only the market to follow our lead, we'll build a better, stronger, and healthier democracy. The high-end consumer can drag the market along with it.

If we make a healthy information diet as normal and obvious as something like a healthy food diet, then those that aren't consuming healthily will begin to feel social pressure. Nobody wants to be ignorant or even have the appearance of ignorance. The social consequences of being seen as ignorant are far more significant than the social consequences of smoking or obesity.

With another divisive election around the corner, I'd like the consequence of you reading this book to not only be your going on an information diet, but also to your starting or joining a local campaign for information dieting with three goals in mind:

1. To increase the digital literacy of our communities with the skills I outlined in Chapter 7: the ability to search, process, filter, and share.

2. To encourage the consumption of local information that's low on our metaphorical trophic pyramid.

3. To economically reward good information providers, and to provide economic consequence for those who provide affirmation over information.

This kind of campaign mustn't revolve around a particular person or personality, but instead be driven from the ground up. As much as I'd like to use the political skills I've learned in the past 10 years to drive a traditional campaign, doing so would go against the principles of the book. Instead, a campaign like this has to be driven at both the geographically and socially local levels: neighborhood by neighborhood and network by network.

Conspiracy in Six Easy Steps

1. **Share this book.** If we're going to do this right, then we need more people to know what a healthy information diet looks like. After you're done with this book, share it with a friend—or, if you're feeling generous (to both the friend and your humble author), buy them their own copy. The principles of digital literacy, humor, attention fitness, and a healthy information diet need to spread if we're going to succeed.

2. **Organize.** There may be an infodiet group in your area. Check out *http://informationdiet.com/local* to see if one exists near you. If not, start a Google Discussion group at *http://groups.google.com*. Name it something that's easily discoverable by people in your community: "InformationDiet Austin" or "InformationDiet East Bay." If you send me a link to your group via Twitter (I'm @cjoh), I'll make sure to link to it on *http://informationdiet.com/local* for other people to find.

3. **Focus and be civil.** In your group, keep the focus on the mission: digital literacy, local information, and changing the economics of your information providers. Your group should practice healthy information diets during your discussions; it's useful to be somewhat strict moderators. Your discussion group should never degenerate into political discussions—that's something that there are plenty of other venues for, and as a group, it's better to steer those discussions to the places where they'd be more appropriate.

4. **Meet. Like, face to face.** Anonymity is useful when speaking truth to power and sparking revolutions, but isn't particularly useful when trying to create civilized discourse. Use *Meetup.com* to find or host regular InfoDiet meetups in your area. Share them with me, too, on *InformationDiet.com* and via Twitter. I'll make sure people know they're happening, and I'll try to attend as many as I can.

5. **Learn.** There's more to this subject than the concepts in this book. If you're looking for things to discuss in your local group, check out some of the great reports that the Knight Commission puts out on the future of information and the media, or read some of the many documents in the further reading and bibliography sections of this book. You can also tune in to the blog on *InformationDiet.com* as more is discovered in the worlds of neuroscience and cognitive psychology.

6. **Act.** The group isn't meaningful unless it causes outcomes useful to its local community. To improve digital literacy in your community, start with kids. Find and fund nonprofits that help to teach these skills to children in local schools or in after-school programs. To get information from the bottom of the trophic pyramid of information, start advocating for your local government organizations (your county and city) to create online data catalogs and make public the data that you're already paying for. The same goes for the media: start demanding that they offer source material rather than provide you with their analysis and perspective.

It's also important to share what you've learned and how you're causing change in your community, to help others that are starting groups in their local communities learn best practices. *InformationDiet.com* has lots of resources to help you, including a discussion board that you can use to connect with other groups across the globe.

The remainder of this book is a call to action for the vast rational conspiracy—ideas and observations that come from my time here in Washington, and my time working with and interviewing civic leaders across the country. It's the empowered information diet: once you lose the fluff and start really seeing what's going on, new priorities arise that require new tactics to accomplish.

The Participation Gap

*"The great lie politicians like me tell people like you is 'vote for me
and I'll solve all your problems.' The truth is, you have the power."*

—Governor Howard Dean[88]

Let's go back to those signs: "Enlist Here to Die for Halliburton" and "Keep
your Government Hands Off My Medicare." For me, they were the signals
that something was wrong with our democracy—that our nation suffers
from an information obesity dilemma, especially in the world of politics.

To figure out what was going on in the information diets of partisan activists
in the United States, I built a simple service that subscribed to the various
political email lists that are sent out by politicians, advocacy groups, and
other political organizations. The result is one of the larger compendiums
of political emails that exists organized by political spectrum. I waded
through the various emails we received, and began to get a taste of what
both sides of the political aisle were talking about.

Figure 11-1 shows what might end up in the inbox of a conservative activist.
Figure 11-2 shows what could end up in the inbox of a liberal one.

---

88 http://books.google.com/books?id=TUjGk7nK1jsC&lpg=PA17&ots=ejqX-emCLg&dq=the%20
biggest%20lie%20people%20like%20me%20tell%20people%20like%20you%20howard%20
dean&pg=PA17#v=onepage&q&f=false

The American Spectator	@2	Inbox	**37 Things You Should Hoard...** - Learn the 37 food items that FEMA "should" b
The Sean Hannity Show	@2	Inbox	**US Poverty Rate Hits Record High** - Hannity Headlines Tuesday, September 1
John Tate	@2	Inbox	**Support the Constitution Day Money Bomb** - Constitution Day Money Bomb De
Daily Events	@2		**Mattera, Buchanan: Attorney General Holder an Insult to D.C.'s 9/11 Memorial** - If you a
Rick Santorum	@2	Inbox	**Stunned** - facebook twitter youtube flickr Billy Did you see last night's debate? I
RNC Research	@2	Inbox	**Broken Promise On Poverty** - Having Trouble Viewing This Email Correctly? V
WND	@2		**The New World Order is not coming – it's already here!** - Email not displaying correctly'
RNC Research	@2	Inbox	**Zombie Tax Hikes Rise Again** - Having Trouble Viewing This Email Correctly?
WND	@2	Inbox	**Evolution theory demolished (by pro-evolution scientists)** - Email not displaying
Newt Gingrich	@2	Inbox	**Media's Love Affair with Obama** - Newt 2012 Dear Friend and Supporter, Reme

Figure 11-1. The potential inbox of a conservative activist.

JB Poersch	@1		**Math** - Democratic Senatorial Campaign Committee Dear friend, "This is not class warfa
Scott Arceneaux, Florida.	@1		**Under Attack** - Florida Democrats Dear Friend, One of the most precious rights in our d
Daniel Mintz, MoveOn.org.	@1	Inbox	**Help tax the rich** - President Obama's "Buffett Rule" would make sure that milli
Media Matters for America	@1	Inbox	**Right-Wing Media Defend The Rich Unless The Rich Person Is Named Warren**
Harry Reid	@1	Inbox	**Callous** - Democratic Senatorial Campaign Committee Dear friend, The GOP o
Guy Cecil	@1	Inbox	**Jaw-dropping** - friend, Have you been watching the Republican debates? Jaw-c
Kathy Kilmer, The Wilder.	@1	Inbox	**Whales, walruses, polar bears need your quick action** - Dear Ruby, We must a
Chip Forrester, TNDP	@1	Inbox	**Mistake** - TNDP Banner Danny, What Republicans refuse to understand is that
The Progress Report	@1	Inbox	**Let Him Die?** - Problems viewing this email? View it in your web browser. The F
Center for American Prog.	@1	Inbox	**The Truth About the House Leadership's Nonjobs Agenda** - Center for America

Figure 11-2. The potential inbox of a liberal activist.

In a world where both sides have information diets like these, our democracy will remain completely paralyzed and divided. Moreover, charged up activists and the organizations they support will drive the public further away from the actual mechanics of power in government.

The "sportsification" of federal politics has made it so we treat elections like athletic rivalries, vilifying the other team at the expense of doing what's right for the country. If this was what motivated your constituents, would you listen to them? As Congress stops listening, people get more furious, building larger megaphones with which to shout at their representatives; and Congress, being unable to decipher what people are saying from the sheer volume of input, simply listens less. It's a destructive loop that causes a great chasm between people and the functions of government designed to listen to them: a participation gap.

The participation gap is the gap between people and the mechanics of power in their governmental bodies. Its cause is our desire to focus on large, emotionally resonant issues over practical problems that can be solved, and the disconnect between what people want out of their government and what it can actually do.

Because of the participation gap, citizens are frustrated and take out that frustration at the polling booth, voting to "throw the bums out" and "elect fresh blood in Washington." As new members of Congress are elected, they must rely on the professional class of Washington—professional staff, lobbyists, and consultants—in order to understand the mechanics of our government. The cycle then repeats itself, our satisfaction with Congress sinks to a new all time low, and we do the same thing, over and over again, expecting a different result: Benjamin Franklin's definition of insanity.

What we never do is look at how to close the participation gap, to more closely connect people with the levers of power in Washington. Instead, we're distracted by issues du jour: the anger at Washington being unaccountable turns into debates over the debt ceiling, healthcare, abortion, guns, or gays. But it never turns into a discussion about how to make the United States government better at representing the interests of those who elect it or solving the great disconnect problem. No matter which side of the aisle you sit on, it's better television to watch pundits talk about polarizing issues than it is to figure out how to make governments work better.

The answer isn't as simple as firing the professional class of Washington, either; banning lobbying in Washington will just create a newly named profession of "citizen activists" that will do the same thing. In order to treat the problem, we've got to figure out what causes it.

The Scalability Problem

The first cause of the participation gap is a problem that technologists would call *scale*. The underlying structures of government aren't designed to handle our present population as it is currently interacting with government.

If you take a look at the Constitution, you'll quickly figure out that the framers couldn't have imagined a union with this many people in it. At our first census, the population of the United States was at less than 10 million people scattered across the 13 colonies that now make up the eastern seaboard. The population of Planet Earth was a measly one billion people. There is no way that the framers could have conceived of a country of 300 million people—roughly a third of the world's population at that time.

But the framers did something smart: they pegged the number of representatives in the lower chamber of Congress, the House of Representatives, to be proportional to population. Our first Congress, in 1789, had 65 members of the House of Representatives for roughly four million people; each member represented approximately 60,000 people.

In 1890, our population had grown, and so did our House of Representatives. Each of the 325 members of the House represented roughly 200,000 people. Then at the turn of the last century, just before our population exploded,

Congress came to the realization that the House of Representatives was getting unruly and incapable of getting anything done, so they put a cap on the number of total representatives that we have: 435.

The result today is a staggering 1:717,000 ratio. The only democratic country in the world with a ratio more unwieldy than ours is India. If you combined the populations of Japan, Canada, Germany, France, Italy, and the UK, and had them all represented by just the number of members of UK's House of Commons, they'd still have a lower ratio than we have in the United States.

It's impossible for one person to accurately represent 717,000 people—it's why candidates have to raise and spend millions of dollars on television advertisements rather than getting to know their constituents. It's why members of Congress have to rely on lobbyists to get ideas on what to do, and it's why the media sensationalizes politics. Today, politicians must come out strong on polarizing issues in order to get the attention of the major media markets they're representing. Thus, it's easier for people to treat Republicans and Democrats like the Red Sox and the Yankees.

Granted, a lot has changed since that 1:60,000 ratio was created: we have gone through the media revolutions of the telegraph, the radio, the television, and the Internet, all of which should positively affect a member's ability to hear from her constituents.

We also have enhanced travel technology and infrastructure: planes, trains, and automobiles, combined with a strong civic infrastructure of roads, highways, train tracks, and airports, make it easier for a member to travel back and forth to hear from her district. But even with this new technology, it's clear that we're dealing with a scalability problem with our democracy.

So how do we solve this problem? The law placing a cap on members of Congress was invented 100 years ago for a good reason: Congress was getting unwieldy. If we reverted to our framers' 1:60,000 ratio, we'd now have over 5,000 members of Congress. It's unlikely they would be able to work as a cohesive legislative body at that level.

Sticking to that ratio would mean rebuilding the Capitol building into something that looked a bit more like RFK Stadium—congresses would look more like trade shows than what we see today—and it would mean a Congress that couldn't effectively get anything done.

It would also be impossible to make happen. Getting two-thirds of Congress to agree to dilute their power to less than 1/200th what it is today seems highly unlikely, and the other way to do it—getting two-thirds of the states to hold a constitutional convention on the issue—seems equally implausible.

Transparency

I started this book with transparency: the idea that governments, corporations, news organizations, and all of our information suppliers start opening up. Transparency is a necessary requirement for the reality-based community. To go on a healthy diet, we need cheap, healthy food options. The critical supply ingredient to a healthy information diet is transparency.

After spending a couple years working on transparency in government, I keep coming back to food. Not only do the changes in agriculture have significant parallels to our information production, but they also teach us a lot about transparency's effectiveness. Data-wise, food is very transparent.

In 1990, the Nutrition Labeling and Education Act was passed by Congress and signed into law by then-president George Bush. The act gave birth to the modern version of what we see on the back of nearly every food item we buy in the grocery store today: the FDA's Nutritional Label.

It's a very sophisticated label. On it, you'll find the caloric content of the food, the vitamins and minerals it contains, and even its ingredients. One can, should one choose to do so, make very effective choices about one's health armed with basic nutritional knowledge and the transparency-focused nutritional labels on food packaging.

But with the labels came the other adjectives. While the nutritional labels are standard on the back of our boxes, the front of our boxes started filling up with meaningless, but great sounding words and phrases, too—words like "all-natural" and "farm raised" and "part of your complete breakfast." The words on the front of the box are so much better than the words on the back. They're brighter, flashier, and much more wholesome sounding than those on the back. Who wants to read about Acesulfame Potassium when you can read about "Real CocaCola Taste and Zero Calories?"

No one would argue that these nutritional labels aren't a good thing and a positive direction for the food industry, but they certainly haven't prevented an obesity crisis. The labeling only affected those with the will to read it and the people willing to understand and constantly study what the words and the numbers actually mean.

In 2008, New York City became the first jurisdiction in the country to require calorie labeling of food sold by restaurants with more than 15 locations. Since the law was passed, a lot of research has been done to determine whether the labeling is having the desired outcome: consumers making more conscious decisions.

Two scientific papers provide a good overview of what's happened. The first, released six months after the law went into effect, measured the choices of lower-income residential neighborhoods and their purchase patterns before and after, at Wendy's, McDonald's, Burger King, and Kentucky Fried Chicken.[89] They then compared the purchase patterns of New York City to those of Newark, NJ, where there are no labeling laws, but the social and economic status of the neighborhood residents are comparable.

The results: more than half of the respondents in New York City said that they saw the calorie counts next to menu items in New York City, and 27.7% of the respondents who saw the calorie counts said that the labels influenced their choices, and more than 10% of respondents reported purchasing fewer calories.

But what did the actual behavior say? People in the locations sampled purchased an average of 825 calories before the nutritional labeling law was passed, and purchased an average of 846 calories after the law was introduced. The difference is statistically insignificant and well within the margin of error. The labels didn't change anything.

Another study shows a different result. This time, instead of targeting low-income families and common brands, researchers from Stanford chose high-income groups shopping at a luxury brand: Starbucks. The researchers, with the cooperation of Starbucks, examined every transaction in 2008 from New York City and Seattle, where the calorie count laws went into effect, and compared them to Boston and Philadelphia, where there were no calorie count laws.

They found that the laws did have an effect. Where the calorie counts were posted, there was a decrease of 6% in calories purchased: from 247 calories to 232. While it's a smaller number than the fast food example, it represents more than twice the percentage difference of that in the fast food study, and is thus more statistically significant. Food calories dropped by 14%, and beverage calories remained unaffected. Further, they found that customers consuming more than 250 calories per trip to Starbucks had even more significant changes in their habits: they consumed fewer calories by upwards of 26 percent. Finally, the researchers found that the effect of the calorie postings was greater in higher income and higher education neighborhoods.[90]

With these two pieces of research, you can draw some interesting conclusions. The first is that people may be willing to make marginal changes. Shaving 15 calories off of your Starbucks purchase may be a more palpable change than shaving 50 calories off of your McDonald's purchase.

---

89 *http://content.ny1.com/pages/downloads/Calorie_study.pdf*
90 *http://www.stanford.edu/~pleslie/calories.pdf*

PART III: SOCIAL OBESITY

The second, and more substantial, conclusion is that maybe this form of transparency only affects wealthier and more educated people who are already trying to be healthy. You can certainly see this in the mostly affluent characteristics of the open government and transparency community.

Even amongst the wealthy, transparency alone won't solve the problem. While it's an improvement, the calorie consumption differences are still negligible in terms of fighting what they're intended to fight: obesity. Even if you shave 15 calories a day, or 5,475 calories a year, off of your diet, you're only talking about losing a little over a pound a year. If managing to reduce your entire 2,000-calorie-a-day diet by 6%, you're still only talking about losing a little over a half-pound per month. While it'd help in the obesity crisis, that alone won't solve any problems.

Yet in the face of the deluge of information and the changing standards of economics, we cling to transparency as a model for increasing our integrity. Writer David Weinberger once claimed, "Transparency is the new objectivity." It's convenient and easy: if our information sources are just transparent about their relative biases, then we'll all be better informed subjects.

Clinging to transparency as a replacement for integrity is a bad idea. The choice between transparency and objectivity is a false choice; what we want is for our journalists, our politicians, and even our athletes to be honest. While both transparency and objectivity are useful tools to draw out that attribute, they're no guarantee that any system or human being will be honest and act ethically.

Let's be glad the folks running our water sanitation facilities didn't listen to Louis Brandeis when he said, "Sunlight is the best of disinfectants," and just leave our water out in the sunlight before recycling it for our use. While he was speaking metaphorically, sunlight is a relatively poor disinfectant or remedy for disease. Ethanol, isopropyl alcohol, bleach—heck, even most of the stuff behind the bartender at your favorite watering hole—all tend to be better disinfectants than sunlight.

More bluntly: if you turn the lights on in a roach-infested apartment, it doesn't kill the roaches, it just makes them organize in the shadows. Sunlight only hides the infestation. To get rid of them, you should clean up the apartment and probably call an exterminator.

Transparency's Dark Side

There is a dark side of transparency. Today, it's a tool used as much by the corrupt and dishonest as it is by those who are actually honest. It's used as an illusion to give the appearance of honesty without the intent of being honest. You can simply claim to be transparent, and create a halo of honesty about you, without actually being honest.

Two factors empower this dark side of transparency. We've discussed them a lot in this book. The first is our deluge of information and facts disguised as entertainment. Even the most open and transparent systems must compete with buckets of information that are more interesting. The second is our poor information diets—that we choose information we want to hear over information that reveals the truth makes the competition all the more difficult.

Whether it is the press, the government, or businesses, without conscious and deliberate consumption, transparency does more harm than good. While it can be used as a means of disinfecting a system, transparency can also be used by the corrupt to create a false association with integrity and honesty. A member of Congress could become a public paragon of honesty and integrity by live-streaming video from his congressional office, yet privately be a crook by selling out America in the coffee shop across the street. When he's caught, he could say, "How dare you question my integrity! I have cameras in my office," and make the prosecution all the more difficult.

What's worse is that Joe Public becomes unwittingly complicit in the crimes created by unethical people being transparent about their dishonesty. If a crime is committed with all the sunlight and electric lights in the world shone upon it, then the responsibility for catching that crime gets, in part, placed at the feet of the public. Transparency in a system lets the real enforcement officials off the hook.

The legend[91] of Kitty Genovese is that she was stabbed to death in broad daylight in New York City, and not a single bystander called for help. The story itself is a bit of a myth (though no less tragic), but serves as an example of how sunlight, the electric light, and calls for help can't stop a crime if a public is either failing to pay attention or unwilling to take a stand. Today, America is much like Ms. Genovese, bleeding to death on the sidewalk while the nation is distracted by partisan rhetoric.

Take *Recovery.gov*, for example. The site was heralded by the Obama administration as an unparalleled view of a huge domestic spending package: the 2009, $787 billion American Reinvestment and Recovery Act. Its function? To combat waste, fraud, and abuse by the system—to ensure

---

91 The social record of Kitty Genovese does not match up with the historical one. Neighbors did turn out to help and did pay attention to her call. Her story is more myth than legend.

that the taxpayer's money was spent wisely and prudently, free from fraud and profiteering.

The Chairman of the Recovery Board, Earl Devaney, said upon the launch of the full *Recovery.gov* website on September 29th, 2009:

> *"I believe that this historic level of transparency will help drive accountability in many new ways. While this Board and countless other federal, state and local oversight agencies will be looking for fraud and waste, every American citizen who clicks on this website has the potential to become what I'm calling a Citizen IG [inspector general]. That's right—we need you to help us identify fraud, waste or mismanagement in your community."*

To date, *Recovery.gov* lists one account of fraud: an incident in South Carolina that ended in the conviction of five people, and saved the taxpayer a paltry two million dollars. Did a "citizen IG" find and report this crime using *Recovery.gov*? No. It came from "on-site agency officials."[92]

With our political information, it's the same thing. We can make all the lobbyist meetings, all the campaign contributions, all the electioneering, every vote, every committee hearing, and every cocktail party open and available, online and in real-time, and even hand deliver them to every person's doorstep—we can even have a giant federal agency label all our info-nutritional information, carefully and ethically. But it's likely to be about as effective as our nutritional labels.

Like the calorie counts from food, transparency is ineffective at arming the masses unless there's a strong will in the public to arm itself with the knowledge of how this information affects us, and how to effectively read the metaphorical labels. People will be no less obese—and no less ignorant—unless they have the will to consume less of the stuff that's bad for them, and more of the stuff that's good for them. While transparency can help the problem, it alone cannot fix it.

Transparency's Potential

That's not to say transparency is always a bad thing. Used in the right way, it is a vital weapon in our arsenal against corruption, just like the nutritional label is a vital tool in our arsenal for a healthy diet. One cannot make healthy food choices without knowing what's in one's food, just as

---

92 *http://www.recovery.gov/News/featured/Pages/Criminal-Convictions-and-Significant-Monetary-Recoveries-for-Fraud-Schemes-Involving-Recovery-Funds.aspx*

one cannot make healthy electoral choices without knowing whom one is voting for, and what kind of influences one has around them.

The other advantage of transparency is that it is, by itself, educational. Dissecting a car engine, and understanding its functions wholly, not only makes you a better mechanic, but is also likely to make you a better driver. Understanding the data that runs your company, and the factors that go into its success, makes you a better employee. And understanding the factors that go into the election of a candidate, and the motivations behind their votes, not only makes you a better citizen, but it also helps you understand government.

The promise of transparency is powerful in the world of government. It begs us to imagine a world in which everyone can see how the government is being influenced—in which our government has little privacy; investigative reporting is made easier because the dots that need connecting come preconnected; and the Pentagon Papers of today divulge themselves, through the miracle of the latest technologies.

But the truth is that citizen-focused transparency initiatives have a miserable track record of fighting corruption. And citizens have a miserable track record of using those initiatives to make rational decisions about the people they elect.

Transparency isn't a replacement for integrity and honesty; it's an infrastructural tool that allows for those attributes to occur—but only if the public is willing act upon the information that they receive as a result of transparency in a conscious, deliberate way. We can't let transparency be a tool for only the rich and well-educated to use to drive their decision-making. It must be woven into our civic fabric, and comprehensible by all.

Bridging the Gap

The answer is to take the problem into our own hands. If we want to maintain our democracy, we've got to solve government's scalability problems, and the way that we solve them is by being active participants in our democracy not just on the second Tuesday of every other November, but on the other 364 days of the year. To solve the scalability problem, we must become active participants in our government.

This doesn't mean becoming a traditional activist. In my political career I've worked on a range of issues—from immigration, to healthcare, to more recently effective government and transparency. There are two big lessons I've learned.

The first is that there's a gigantic gap between the skills it takes to win an election and the skills it takes to govern a country. It turns out that electing people—the skills of people like David Axelrod and Karl Rove—are advanced, learned skills that require years of experience to get right.

The skills that it takes to persuade you to vote for someone are entirely different skills than the skills it takes to run a country. Managing the world's largest budget, determining how the government can buy things, figuring out how to take and use public comments—from the soldier in the army to the head of the White House's Office of Management and Budget—they're all skills that are necessary for the management of our government, and they're not political skills, they're governmental ones.

Yet all of our activism pours into the former skill, and none to the latter. If we really want to fix our government, we've got to be participants in the way government works, not who it employs.

The second lesson I learned is that many of the nonprofits and advocacy groups are more interested in staying relevant than solving problems. The motives of many advocacy organizations are not to solve the issues they're working on, but rather to continue to raise money and make payroll. As a result, these advocacy groups tend to focus on larger problems that can go unsolved for years.

While some of these organizations do important work—pushing the envelope to make the United States healthier, to make the environment cleaner and more sustainable, or to try and increase the effectiveness of education in the United States—there are smaller, non-partisan battles that could be won that could have long-standing benefits to the country.

Presuming that our government isn't going anywhere, what can we do to make it better?

Since the "great experiment" started, America's weakness, as de Tocqueville noted, was and still is the tyranny of the majority. My plea to you is to start sweating the small stuff at the expense of some of the big stuff. Washington isn't the land of vast, radical changes, it's a battleship waiting to be nudged in the right direction. Let the legions of information-obese fight on the front lines, and join me in nudging the small nuts and bolts that hold the ship together.

If you're worried about federal spending and the budget, don't concern yourself over the debt-ceiling debate. Work to change procurement laws so that government can get access to the same things the private sector has without paying an arm and a leg. We spend so much time figuring out what programs to spend money on, comparing their priorities to one another, and blanket cutting them when they're deemed too luxurious. It's the equivalent of trying to lose weight by cutting off your legs. Optimizing how government spends its money is at least as important as figuring out what our money gets spent on, and there are real, pragmatic solutions to getting there.

If you're interested in making government more accountable, work on making it so that the government's listening tools and policies are modernized. Many government agencies have legal teams that feel as though social media is an appropriate place for its communications team to publish press releases, but not an appropriate place to solicit real comments for regulations. It's mainly because of identity issues: the government wants a physical address, and doesn't trust that a social media profile is fraud-proof. I'd suggest that it's just as easy to lie about your address as it is a social media profile.

Today, the feedback you give an agency through a website like Facebook or Twitter stops at the new media team inside the agency, and never gets involved in the regulatory process. If the Department of Energy is to publish press releases and invite people to interact with its communications department, it also needs to be able to legally take feedback for the regulations it proposes. It's a simple, nonpartisan problem that could be fixed with a few hundred people demanding that the government use the Internet to be real, active participants with us.

If you're worried about Congress being manipulated by money, the United States House of Representatives started filing their campaign contributions electronically a decade ago, yet the United States Senate refuses to do so. Year after year a bill is proposed, and one way or another it ends up suffocating and dying by the end of the session. This results in a half-million dollar expense to the taxpayer as the Federal Election Commission takes nearly three months to type in, from the various campaigns' paper reports, every campaign contribution that every Senate campaign receives. And as a result, we cannot see how a member of the United States Senate is being influenced by money until long after the time when the relevance of that information has passed.

If you're worried about prisons and civil rights, or making America innovative again, take note of the fact that our laws are generally distributed and archived by for-profit corporations, making it so that even access to the laws that we must follow are behind paywalls. Federally funded scientific research also sits in archives only available to those that agree to pay twice (once with their tax dollars, once for the access) for it.

These are small, solvable problems that don't require millions of dollars or people to fix: they require thousands of focused, smart people to push the right levers inside of the government.

We can also improve our government without waiting on government to act. Organizations like PopVox.com, for instance, make it easier for people to translate what they want their representative to do into the language our representatives speak. There's a whole world of technology out there waiting to be used to help members listen to their constituents, and it's likely—now that much of our discussions about politics are public—that we don't need government to act: we can build tools that listen to what people are already saying, make that information public, and question our elected officials when they're voting against their constituencies.

At the local level, there are thousands of opportunities and willing participants on the side of government. *SeeClickFix.com,* for instance, builds tools that integrate with various cities' request hotlines so when a citizen spots a problem—say a pothole—they can easily report it back to the government. And more importantly, if they spot something they or a group of people can fix themselves—like picking up litter in a park—they can use the site to organize people to help pick up the litter.

These are just examples of what I'd like to believe Governor Dean meant when he said, "You have the power." We mustn't rely on our government alone to solve our problems for us. We have the ability to do it for our neighbors, our communities, and our country as a whole.

Every issue—healthcare, the environment, immigration, even defense— has hundreds of small, nonpolitical, operational problems waiting for a solution, and fixing these small things can have a huge impact compared to combatting a vague foreverwar on issues built to perpetuate the system of donor dollars, consultants, and lobbyists.

The trick is the information diet: filtering out the nonsense meant to get us charged up on issues that will take years to solve, and becoming educated and smart about our government. If we want our government to change, we have to start taking responsibility for not just electing new people, or passing big policies, but sweating the small stuff too.

Political Infoveganism

The rule at most dinner parties is that there are three things you don't discuss: sports, religion, and politics. I can understand the first two—no sporting event is useful without an intense rivalry, one that's built intentionally to cross the wall of logic and rational behavior and into something more akin to faux-tribal loyalties. And religion is a deeply personal belief that's usually nonpliable. It's likely too difficult to get a Muslim and a Christian to agree on the stature of Jesus Christ or Muhammed.

Politics are different. The greatest political ideas have come from the constant search for synthesis and pragmatism, and the foundation of democracy is constant public participation. Policy is something we should talk about at the dinner table; it's vital to our civic health that we do. Democracy cannot survive without the synthesis of ideas from its citizens.

Yet the reason we don't is because we risk relationships when we do. It is because of the fear of the Uncle Warren situation: that the conversation will devolve from ideas to attacks, name-calling, and finally to division. It's not worth the risk. Bringing up politics always ends up with alienation.

The source of our problem with political dialog has its roots in our information diets. Frequently, mainstream national political news is worthless—at best it glosses over the issues that governments are trying to deal with, and at worst represents sensationalized opinion. From Dylan Ratigan to Bill O'Reilly to Wolf Blitzer, paid political operatives and pundits gloss over the facts in order to keep you watching. As someone who has worked inside D.C.'s machinery for a decade, I have learned that the media class around the United States Federal Government and national news has little interest in providing you with the public service of informing you. They are interested in selling advertisements.

Our political information diets are the worst of them all—they're misinformed, they offer little to no knowledge about the actual procedures of Washington, and deliver to us the news we want to hear, not the news we should hear. As a result, we grow more attached to the teams of our choosing—the reds vs. the blues, rather than finding the great synthesis of ideas.

Political news does us no good unless it is potentially actionable via our votes or our activism. To make sense of politics, we need to delve underneath what our news outlets are telling us and into the data that makes politics tick. Thanks to the work of organizations like the Center for Responsive Politics, the Participatory Politics Foundation, and my former employer the Sunlight Foundation, we can start having a direct relationship with what's really happening in the White House and on Capitol Hill.

A healthy information diet always starts locally—and your political information should be no different. The goings-on of your state representatives and city and county governments, along with your school boards and other local government offices are the best, healthiest forms of content for political news, and should be consumed over the national or global news.

OpenGovernment.org, a project of the Participatory Politics Foundation and the Sunlight Foundation, is attempting to build user-friendly websites that allow you to see every vote cast on every bill by every elected state representative and senator in every state. It's a huge undertaking, and if you're in one of the states they're covering, then you can take advantage of a great user interface and user-focused thought that goes into the project.

At the local level, the National Institute for Money in State Politics tracks non-federal races: your governors and state representatives. With its website *followthemoney.org*, you can type in the name of a politician and see who is funding his campaign. At the federal level, the Center for Responsive Politics' *opensecrets.org* does the same thing.

At the federal level, *OpenCongress.org*, also a project of the Participatory Politics Foundation, gives you unprecedented access to what's going on in the United States Senate and House of Representatives. Just like *OpenGovernment.org*, you can find your politicians, look up their votes on bills, and even contact them to tell them how you feel about issues.

To see what influences our politicians, it's good to take a look at the industries and donors giving to their campaigns. While the connection between money and politics isn't direct and uniform, money at the very least buys access, and at its worst buys votes. In either case, it is good to take a look at what kinds of people you're associating with by supporting a particular candidate or campaign.

On television, C-SPAN does a better service of covering the news than FOX, MSNBC, or the major networks. It provides an advertisement and analysis free way for you to see what's going on and to see what candidates are saying directly.

For activists, it may seem nice to subscribe to political emails too, to get the latest on the campaigns and issues that you support, but most of the time, these too quickly turn into advertisements. Sign up to get updates from Newt Gingrich, for instance, through *HumanEvents.com*, and you'll soon start getting emails asking you to buy gold, advertisements for books to read, and recommendations on penny stocks.

Finally, to keep your inbox from filling up with political advertisements, avoid signing petitions and signing up for regular campaign updates. As a cofounder of one of the larger firms on the left responsible for the drafting of these petitions and the software that runs it, I can assure you that the online petitions that you sign are not meant, primarily, to cause change. They're meant to get your email address so that you can later be bombarded by emails asking for money.

Instead, keep your voice your own, and if there's an issue that you care about, bypass the middlemen and speak directly to your representative through the official means given to you—via *house.gov*, *senate.gov*, or *whitehouse.gov*. Or if you want to be truly effective, meet with your representatives in person. Call their offices, ask to speak to their schedulers, and get yourself a meeting.

With business news, paying attention to your local businesses, reading the public filings of companies from the SEC is likely to give you more benefit than listening to Jim Cramer smash things on CNBC.

In sports, developing a mastery of the statistics we use to measure the performance of our athletes may provide you with more insight and more pleasure for the game than listening to the washed up pundits and armchair quarterbacks tell you what they think. And certainly watching the games themselves is far more important to understanding the game than listening to the pundits prattle on about it.

It turns out the more local your sports diet, the more rewarding it can be too. Although watching a local high school baseball game doesn't often give us the athletic showmanship of professional sports, it trades that for being able to watch kids play for the sport of the game, rather than for the money.

The same can be said for any major section of your newspaper, or any topic you're interested in. The pattern here is simple: seek to get information directly from sources, and when the information requires you to act, interact directly with those sources. An over-reliance on third party sources for information and action reduces your ability to know the truth about what's happening, and dilutes your ability to cause change.

The thing that's made what Alexis de Tocqueville called "The Great American Experiment," as on page 135, work is our ability to be pragmatic. Unfortunately, the economics of our information production, and what we're willing to consume, is destroying our very ability to be pragmatic—to look to solve solvable problems. We get caught up in big debates, and brush off the boring stuff for the wonks to deal with.

Going on a healthy information diet restores our ability to be pragmatic. Let's take our country back, not from the right or from the left, but from the crazy partisanship of both sides. Let's give it to the stewards that have made the country so great, the pragmatists—the ones who want to create a more perfect union. A country with measurable results and demonstrably good outcomes.

Without stealing too much from President Obama, I'd like to suggest that we are the wonks we've been waiting for.

Dear Programmer

*"It circulates intelligence of a commercial, political, intellectual,
and private nature, with incredible speed and regularity. It thus
administers, in a very high degree, to the comfort, the interests, and
the necessities of persons, in every rank and station of life. It brings
the most distant places and persons, as it were, in contact with each
other; and thus softens the anxieties, increases the enjoyments, and
cheers the solitude of millions of hearts."*

—Supreme Court Justice Joseph Story in 1833
on the United States Post Office[93]

Six thousand years ago, there was a professional class of people that had a
better relationship with information than everybody else. The professional
scribe, armed with the ability to read and write, had a better ability to
figure out the world than anybody else. Scribes became more than just
stenographers for the courtrooms of power; they explored the sciences,
becoming mathematicians, scientists, architects, and physicians. For
millennia, the scribe wasn't just a professional class, it was the backbone
of civilization.

Through the development of the printing press, and a global push for basic
literacy, the scribe class became obsolete. Knowing how to read and write
wasn't a trade secret for a professional class—it was a necessary asset
for economic survival. Scribes went extinct, and were replaced in society
by journalists, who had marginally better abilities to read and write, to
preserve the link between the people and the truth.

But our romantic idea of the journalist speaking truth to power has now
gone all but extinct. As our media companies have consolidated and sought
shareholder returns over civic responsibility, there's not much left for the

---

93 Joseph Story, Commentaries on the Constitution 3:§§ 1119–42, 1144–45.

investigative reporter; local newspapers just don't have the budget for investigative reporting, and larger media companies are making too much money peddling affirmation over information.

The invention of the printing press brought with it the Protestant Reformation—a democratization of the people's relationship with God. Once the Bible could be purchased by the middle class, every man, in the eyes of Martin Luther, could become his own priest. Today, the invention of the Internet has democratized information such that professional journalists alone cannot own the relationship with the facts anymore.

Today, programmers are the new scribes. Whether it is the developers at Google, determining which search results are accurate for a particular query; the developers at Microsoft, building the browser that most of us use; the developers at Apple, building the latest phones so that we can have a printing press in our pockets; or the developers at Facebook, figuring out which of our friends are the most relevant to us—the developers build the lenses that the rest of us look through to get our information.

This book's agenda is to help people make more sense of the world around them by encouraging them to tune into the things that matter most and to tune out the things that make them sick. The ones who can link the public with the truth most effectively today aren't journalists, they're developers. As the digital divide continues to close, and as a generation of children grows up knowing how to use an iPad from the age of two, developers must take the mantle of scribe seriously and responsibly.

The opportunities for developers to make a difference are unparalleled. The self-driving cars being engineered at companies like Volkswagen and Google aren't just novel inventions that allow us to watch movies on our way to work; they're life-saving devices. The self-driving car promises a future in which drunk driving deaths no longer happen.

The World Bank has opened most of its data to the public, hoping that developers can find more effective ways for the organization to distribute financial and medicinal aid to developing nations.

Code for America is creating an army of developers to create technology that helps the government provide cheaper, more transparent, and more reliable services. In its first year, it managed to create new ways for civic leaders to work with one another in Philadelphia and Seattle, and provided more educational transparency to the city of Boston. Through its Civic Commons project, it's helping municipalities work together to lower the costs of the software they procure by connecting the cities together to share.

Just after the devastating earthquakes in 2010, I hosted a "Hack for Haiti" event at the Sunlight Foundation. In just 48 hours, a small group of developers at a company in Washington called Intredia developed software that allowed relief workers on the ground to translate Creole into English without the need for an Internet connection.

Most developers haven't taken this new responsibility to heart. A half-century ago, the brightest minds of the generation were working on putting a man on the moon. Today, the 20-something research scientist and data team lead for Facebook, Jeff Hammerbacher, put it best: "The best minds of my generation are thinking about how to make people click ads."[94]

If you're a developer, you can do more than this: you can solve problems. With the right data, and working with the right people, you can find efficient ways to connect vaccines with the people who need them the most, and prevent them from being wasted on the people who need them the least. You can find ways to close the gap between the reality-based community, and the folks stuck in epistemic loops, by linking them more closely to the levers of power in their community.

My plea to you is that you take your role in society seriously. Find an issue you care about: the environment, cancer, space exploration, education, rewiring communities, pet adoption—anything—and dedicate some portion of your time to finding new ways to put your skills to use in that community.

You needn't ask for permission to do this. Do not wait for a nonprofit or advocacy group to ask you donate your time. While it's useful to partner with organizations, it's likely that they're more interested in your skills to help them fundraise than they are to solve problems. Instead, find ways to interview and understand experts in the field, and then invent new ways to solve problems big and small. The best ideas do not rely on a government's or organization's permission or compliance for implementation. The best ideas provide irrefutable insight and solve problems.

The lean startup world that many technology-focused people find themselves in usually starts with a business-oriented cofounder, and a technology-oriented cofounder. To make an interesting social contribution, try partnering up with a journalist. Cynicism aside, there are still a few good reporters working in the world, who know how to ask the right questions and get the most out of the data that you can process.

There are networks of journalists looking for developers across the country. Check out the organization Hacks/Hackers, which is attempting to do just

---

94 *http://www.smh.com.au/business/world-business/why-this-tech-bubble-is-different-20110415-1dhbm.html#ixzz1YEPeAxNW*

that: link great developers with great investigative reporters to combine the best of both worlds. Watch the work of the Knight Foundation, too. They're investing millions of dollars in reinventing media for the digital age.

Keep in mind that this isn't a call for you to build apps for your favorite nonprofit. Unless you're willing to support and maintain each application, and help constantly ensure its usage and adoption, you're wasting your time. Your nonprofit likely doesn't have the kind of resources or knowledge it takes to ensure success. Rather, it's a call for you to solve problems using your skills.

Doctors Without Borders works because doctors can triage the ill, and put them on a course to getting better. They're solving immediate problems, and when they leave, the doctors know they made a difference. A programmer's relationship to her product is different: it takes time and maintenance to have the desired effect.

My other plea to you is that you take your role in society responsibly. Just as responsible journalists have a code of ethics, so should you. It should never be your goal to analyze data to make a point, but rather to analyze it to tell the truth. As we've discovered in this book, we all come with our own biases—some we don't even know we have. But you must try as hard as you can to not let your own agenda supersede the truth.

The CEO of my publisher, O'Reilly Media, Tim O'Reilly, has a guiding principle that I think applies here: work on stuff that matters. Please, don't let your entire career be about figuring out new ways to deliver advertisements. Even if it pays the bills, find an additional outlet to use your skills to make a difference.

The greatest scribes have always done so, whether it was Imhotep and the construction of the Pyramid of Djoser, Martin Luther and his 95 Theses, or Google's self-driving car, our information technology is powerful stuff. You can do amazing things if you, as O'Reilly says, take the long view, and create more value than you capture.

You can even run for Congress. While many sneer at the idea of the nerdy caricatures of developers that they know, the fact is that software engineers are often great communicators. And while cynical developers may be repulsed at the idea of working for such an organization, there's so much value they can add.

Developers are great at using technology to connect directly with people in ways that others cannot, and at helping constituents connect with one another. With a developer who understands the guts of the Web in a leadership spot inside Congress, Congress could start communicating more effectively online. And as this developer became more successful, the rest of Congress may very well follow suit.

The government's problems are becoming increasingly technical—or the problems we're facing have technology tied to them in some way. For instance, the American Recovery and Reinvestment Act of 2009 isn't just a 1000+ page bill that's now a law, it's also a technical specification for *Recovery.gov*—and it's written by people who don't know how to write specifications. Worse, unlike a poorly informed client or boss, if you don't adhere to this client's wishes, you don't just lose money—you may be breaking the law. Thus, Recovery.gov was built to spec, but hasn't been particularly effective at bringing people into the process.

It's every crooked consultant's dream to have a client who views what they sell as a form of mysticism, and that's precisely what's happening around our muncipal, state, and federal governments. A few developers in Congress could reign in the spending and help their peer representatives appropriate taxpayer funds. Today, there is exactly one developer who has written software professionally who has also been elected to Congress: Rep. Steve Scalise of Louisana. If a revitalization of government technology is going to happen smartly and wisely, we need some developers inside Congress to help lead the way.

Of course, you don't have to (and probably shouldn't) start in federal politics. Join your local civic association first, and find new ways to help your local community. You'll discover plenty of opportunity and many open arms there. But again, don't wait for someone's permission, unless it absolutely requires their adoption and sponsorship in order to work.

Finally, for those of you who aren't engineers, know that the most vital thing after basic literacy for the education of yourself and your children is digital literacy and STEM education: Science, Technology, Engineering, and Math. History shows us that perhaps a century from now saying "I'm not an Internet person" may be much like saying "I don't know how to read." Organizations like CodeNow are helping transform our concepts of literacy by making sure computer science education is accessible to everyone who wants it, and is constantly looking for volunteer engineers who can help teach classes. While it's not the key to solving all of our problems and differences, those skills, combined with the ability to communicate, give us the greatest ability to see the truth.

Further Reading

People

The concept of an information diet is a relatively new one, and the thoughts and ideas in this book come from research and interviews with scores of people. In addition to pointing you towards the research papers and books I've read and recommend to further your study, it's also important to follow the people who are leading this field, who are studying how the mind works, the economics of information, and the ever-changing face of our news media.

As much of our scientific research still sits behind paywalls, interacting directly with the scientists who use social media has an added payoff: you'll gain exposure to their work without having to subscribe to the various scientific publication services. In the spirit of infoveganism, I advise you to connect directly with these researchers and scientists.

Matt Cutts

Matt Cutts is the head of the web-spam team at Google, the person with the job of managing Panda, and maintaining Google's delicate search relationship with content farms. He's been called "Google's Greenspan."

http://twitter.com/mattcutts

http://mattcutts.com

Marco Iacaboni

Dr. Iacaboni's insight on the consequences of neuroplasticity and how we affect each other is tremendously important to follow.

http://twitter.com/marcoiacoboni

http://iacoboni.bmap.ucla.edu/

Ryota Kanai

Dr. Kanai's research links our brain's structure to our political affiliations. His continued interests are around our perception of time, the neuroscience behind our attention, and distractibility.

http://twitter.com/kanair

http://www.icn.ucl.ac.uk/Research-Groups/awareness-group/group-members/MemberDetails.php?Title=Dr&FirstName=Ryota&LastName=Kanai

Brendan Nyhan

Dr. Nyhan's work on measuring the effectiveness of messaging on the public and the outcomes of our information consumption is leading the field. Read his papers and engage with him online. He's responsive and smart.

http://twitter.com/BrendanNyhan

http://www.brendan-nyhan.com

Robert Proctor

Robert Proctor invented the term *agnotology*, and was the inspiration for Chapter 7.

http://www.stanford.edu/dept/HPS/proctor.html

Julian Sanchez

Julian Sanchez is the person who brought the idea of epistemic closure into the modern political dialog. He's a writer for Reason magazine and the CATO institute.

http://twitter.com/normative

Linda Stone

Linda Stone's research on conscious computing, email apnea, and our attention spans is amazing to watch. Follow her work at:

https://twitter.com/LindaStone

http://lindastone.net

John Tierney

John Tierney is a science columnist for the *New York Times* and, along with Roy Baumeister, is the author of *Willpower: Rediscovering the Greatest Human Strength.*

http://twitter.com/JohnTierneyNYT

http://topics.nytimes.com/top/news/science/columns/johntierney/index.html

Jeff Jarvis

Jeff Jarvis is an associate professor and director of the interactive journalism program and the new business models for news project at the City University of New York's Gradute School of Journalism. While he's not directly quoted in this book, his thought leadership around the field of journalism is worth paying attention to.

http://twitter.com/JeffJarvis

http://buzzmachine.com

Dan Gillmor

Dan Gillmor teaches digital media entrepreneurship and is founding director of the Knight Center on Digital Media Entrepreneurship at Arizona State's Walter Cronkite School of Journalism and Mass Communication. His work is trying to make sense of the new fields of journalism and how the digital industrialization of it can yield new business models.

http://twitter.com/dangillmor

http://dangillmor.com

Jim Gilliam

Jim Gilliam is the consummate civic hacker, using his skills to try and connect people to each other and to the levers of power in their local communities. He's the founder of 3dna, a startup in California that builds tools to shake up a political system, most recently NationBuilder, an affordable tool that allows people to organize effectively.

http://twitter.com/jgilliam

http://3dna.us/blog

And of course, me

My hope is that this book isn't the end of something, but the start of something: an exploration into how our information affects our health. I certainly have not provided all the answers in this book, and there's still so much more work to be done. Please, be in touch.

You can find me on Google+ at: *http://gplus.to/cjoh*

I use Google+ to have discussions with people about the topics of this book, and to do the occasional video chat with people interested in the field. Please stop by and interact with me.

Twitter: *http://twitter.com/cjoh*

I tend to use Twitter to broadcast my latest writing, and to share simple links about the field of information dieting, government data, and activism.

Books

Behind this book lies scores of others, and I've drawn from the research of many others to write this one. If you'd like to pursue studying in this field, I recommend the following:

Ariely, Dan. *Predictably Irrational: The Hidden Forces That Shape Our Decisions.* Harper Perennial, 2010.

Baumeister, Roy F., and John Tierney. *Willpower: Rediscovering the Greatest Human Strength.* Penguin Press, 2011.

Brown, Stuart, M.D., and Christopher Vaughan. *Play: How It Shapes the Brain, Opens the Imagination, and Invigorates the Soul.* Avery, 2009.

Carr, Nicholas. *The Shallows: What the Internet Is Doing to Our Brains.* W.W. Norton & Co., 2011.

Hurley, Matthew M., Daniel C. Dennett, Reginald B. Adams Jr. *Inside Jokes: Using Humor to Reverse-Engineer the Mind.* MIT Press, 2011.

Manjoo, Farhad. *True Enough: Learning to Live in a Post-Fact Society.* Wiley, 2008.

Medina, John. *Brain Rules: 12 Principles for Surviving and Thriving at Work, Home, and School.* Pear Press, 2009.

Pariser, Eli. *The Filter Bubble: What the Internet Is Hiding From You.* Penguin Press, 2011.

Putnam, Robert. *Bowling Alone: The Collapse and Revival of American Community*. Touchstone Books, 2001.

Scully, Matthew. *Dominion: The Power of Man, the Suffering of Animals, and the Call to Mercy*. St. Martin's Griffin, 2003.

Shirky, Clay. *Cognitive Surplus: How Technology Makes Consumers into Collaborators*. Penguin Press, 2011.

Stoll, Clifford. *Silicon Snake Oil: Second Thoughts on the Information Highway*. Anchor, 1996.

Tavris, Caroll, and Elliot Aronson. *Mistakes Were Made (But Not by Me): Why We Justify Foolish Beliefs, Bad Decisions, and Hurtful Acts*. Mariner Books, 2008.

Vaidhyanathan, Siva. *The Googlization of Everything (And Why We Should Worry)*. University of California Press, 2009.

Blogs

Three blogs kept me informed on the latest developments of science around this field, and their work is really remarkable. I encourage you to be a regular reader of each, if you want to constantly fine-tune your information diet. The debate in the science community over what I've discussed in my book is lively, and worth paying attention to.

Deric Bownds' Mindblog: *http://mindblog.dericbownds.net/*

Jonah Lehrer's Frontal Cortex: *http://www.wired.com/wiredscience/frontal-cortex/*

Re:Cognition: *http://thebeautifulbrain.com/category/recognition/*